EXPLAINING TECHNICAL CHANGE

STUDIES IN RATIONALITY AND SOCIAL CHANGE

STUDIES IN RATIONALITY AND SOCIAL CHANGE

Editors: Jon Elster and Gudmund Hernes

Editorial Board:
Fredrik Barth
Amartya Sen
Arthur Stinchcombe
Amos Tversky
Bernard Williams

JON ELSTER Explaining technical change
JON ELSTER & AANUND HYLLAND (EDS.) Foundations of social choice theory
JON ELSTER (ED.) The multiple self
JAMES S. COLEMAN Individual interests and collective action

Jon Elster

EXPLAINING TECHNICAL CHANGE

A Case Study in
the Philosophy of Science

Universitetsforlaget
Oslo   Bergen   Stravanger   Tromsø

Published in collaboration with Maison des Sciences de l'Homme, Paris

CAMBRIDGE UNIVERSITY PRESS
Cambridge, New York, Melbourne, Madrid, Cape Town, Singapore, São Paulo, Delhi

Cambridge University Press
The Edinburgh Building, Cambridge CB2 8RU, UK

Published in the United States of America by Cambridge University Press, New York

www.cambridge.org
Information on this title: www.cambridge.org/9780521270724

© Cambridge University Press and Universitetsforlaget 1983

This publication is in copyright. Subject to statutory exception
and to the provisions of relevant collective licensing agreements,
no reproduction of any part may take place without the written
permission of Cambridge University Press.

First published 1983
Reprinted 1985, 1988, 1990, 1993
Re-issued in this digitally printed version 2009

A catalogue record for this publication is available from the British Library

Library of Congress Catalogue Card Number: 82-9702

ISBN 978-0-521-24920-1 hardback
ISBN 978-0-521-27072-4 paperback

Contents

PREFACE AND ACKNOWLEDGEMENTS 7
GENERAL INTRODUCTION 9

PART I: MODES OF SCIENTIFIC EXPLANATION
INTRODUCTION TO PART I 15
1. CAUSAL EXPLANATION 25
    a. Causal explanation: in general 25
    b. Causal explanation: in the social sciences 32
2. FUNCTIONAL EXPLANATION 49
    a. Functional explanation: in biology 49
    b. Functional explanation: in the social sciences 55
3. INTENTIONAL EXPLANATION 69
    a. Intentionality 70
    b. Intentionality and rationality 72
    c. Rationality and optimality 74
    d. Intentionality and causality 83

PART II: THEORIES OF TECHNICAL CHANGE
INTRODUCTION TO PART II 91
4. NEOCLASSICAL THEORIES 96
    a. The production function 96
    b. Explaining the factor-bias of technical change 101
    c. Explaining the rate of technical change 105
5. SCHUMPETER'S THEORY 112
    a. *The Theory of Capitalist Development* (1911) 113
    b. *Business Cycles* (1939) 120
    c. *Capitalism, Socialism and Democracy* (1942) 125
6. EVOLUTIONARY THEORIES 131
    a. Animal tool behaviour 131

b. Eilert Sundt   135
    c. Nelson and Winter   138
    d. Paul David   150
7. MARXIST THEORIES   158
    a. Did Marx believe in fixed coefficients of production?   159
    b. The rate and direction of technical change   166
    c. The falling rate of profit   178
    d. The development of the productive forces   181

Appendix 1: RISK, UNCERTAINTY AND NUCLEAR POWER   185
Appendix 2: THE CONTRADICTION BETWEEN THE FORCES AND RELATIONS OF PRODUCTION.
    With a Mathematical Note by Aanund Hylland   209

NOTES   237
REFERENCES   259
INDEX   271

Preface and acknowledgements

The occasion for writing this book was provided by Bernt Schiller of the University of Linköping (Sweden), who asked me to write a textbook in the philosophy of science that could be suitable for their doctoral programme 'Technology and social change'. I am grateful for the suggestions and comments offered by him and his colleagues along the way. I should also like to thank the following for their comments on an earlier draft: G. A. Cohen, Aanund Hylland, Michael MacPherson, Nathan Rosenberg, and an anonymous referee of Cambridge University Press. Acknowledgements for comments on the Appendices are given at the appropriate places. Appendix 1 was originally published in *Social Science Information* 18 (1979).

Part I of the work can be read as an introduction to the philosophy of scientific explanation. The Achilles heel of this part, clearly, is the chapter on causal explanation. My competence in these intricate matters is not high, but since the chapter was required by the overall architectonics of the book I felt I should state my views even when they are not strongly grounded. I hope I have avoided saying too much that is obviously wrong, but the reader may justifiably feel that some of what I say is not very interesting or not as tightly argued as he might wish.

Part II is *not* to be read as an introduction to the theory of technical change. It is subordinated to the epistemological purpose of showing how the distinctions and propositions of Part I can be applied to a specific set of empirical problems. Here the danger is that my exposition of the theories may be too compact for the non-specialist and too sloppy for the specialist. To the first, I can only offer the advice to look up the original works. To the second I make a plea that ambiguous statements be taken in their most plausible sense. Even so, there will probably remain some statements that are plainly wrong, of which the reader can justly complain.

J.E.

General introduction

The study of technical change is uniquely well suited to epistemological analysis. It is located at the interface of social science and the natural sciences, and so might be expected to be relevant to the discussion of 'the unity of science'. It bridges the gap between pure science and everyday affairs, and might therefore be expected to throw light on how theoretical knowledge relates to the observable world. Technical change – the manufacture and modification of tools – may have played an important role in the evolution of intelligent life on earth, comparable to that of language. During the course of human history, social institutions have emerged and disappeared largely in response to changes in productive and destructive technology. Moreover, technical change offers a challenge to analysis in that it is fundamentally unpredictable. 'If I knew where jazz was going, I'd be there already', Humphrey Lyttelton is reported to have said. Similarly, any attempt to explain technical change sooner or later comes up against the paradox of turning creativity into a dependent variable.

In this book I first set out the main varieties of scientific explanation, and then look at some central theories of technical change from the vantage point provided by that discussion. This enables me to deal with what are, I think, the two main approaches to technical change. First, technical change may be conceived of as a rational goal-directed activity, as the choice of the best innovation among a set of feasible changes. Secondly, technical change may be seen as a process of trial and error, as the cumulative addition of small and largely random modifications of the production process. Any serious student of technology will agree that technical change exhibits both these aspects, but there are strong differences in emphasis between the contending explanations.

This main dichotomy cuts across many of the other relevant distinctions that can be made in this domain. I believe, for instance, that neoclassical and Marxist theories of technical change at the level of the firm share an emphasis on the rational-actor approach. True, Marxists have not engaged in much detailed modelling of technical change at the micro-level, but there are a number of historical studies by Marxists who argue that the entrepreneur uses innovation as a weapon in the class struggle. The neoclassical economists explain technical change in the light of profit maximization, whereas Marxists tend to argue that power rather than short-term profits is at stake. Within both traditions technical change is explained in the light of the goal to be achieved, although they impute different goals to the entrepreneur.

On the other side of the main dichotomy we find the 'evolutionary' theories of technical change, which emphasize past history rather than future goals in the explanation of why firms currently use the techniques they do. Typically the proponents of these theories look at technical change as more or less closely analogous to evolution by natural selection. It is instructive, therefore, to consider the socio-biological studies of animal tool behaviour which explain technical progress as a literal rather than a metaphorical instance of biological evolution. These theories set out to explain not only specific inventions, but also the emergence of genes for inventiveness. Interestingly, tool behaviour turns out to be closely related to play behaviour – a reminder that creativity is of the essence in technical change.

Technical change may be studied at various levels of aggregation, and for various time spans. Neoclassical and evolutionary theories tend to study change at the levels of the firm and the industry, in contradistinction to the large-scale historical syntheses offered by Schumpeter and Marx. Schumpeter probably is the most influential single writer on technical change, its causes and consequences. Again, this may be because he emphasized creativity and disequilibrium, rather than trying to fit technical change into the pattern of routine profit maximization. He praised capitalism not because of its efficiency and rationality, but because of its dynamic character – to be explained in terms of irrational expectations and dreams of founding private dynasties. Marx also insisted on the uniqueness of capitalism in that, in contrast to all earlier modes of production, it does not oppose technical change, but rather depends on it. Yet in his materialist conception of history he also, somewhat inconsistently, argued that the development of the productive forces is the major determinant of social change in all modes of production.

It ought to go without saying that I am not offering an introduction to theories of technical change. In most cases, the reader will have to go elsewhere for detailed expositions of the various theories discussed in Part II. The discussions of Marxism in Ch. 7 and in Appendix 2 are somewhat more detailed, because I happen to know more about this tradition than about the others. But even in this case my attention is directed by epistemological concerns rather than by an informed interest in the substantive issues. This is a case study in the philosophy of science, not a bird's eye view of science. Again, it ought to go without saying that I am not out to tell economists or historians how to do their work. I believe, however, that philosophers of science can be of help in distinguishing true from spurious foci of disagreement within the empirical disciplines. Empirical work conducted in isolation from the philosophy of science may be no worse for that, whereas the philosophy of science atrophies if it is not in close and constant touch with the development of current thinking on empirical matters. Yet the asymmetry is not so radical as to make philosophy of science totally parasitic.

This will be admitted by many scientists with respect to problems of verification and falsification. Few will contest the statement that Popper's methodology of science – basically an injunction to scientists to stick their necks out – has had a valuable influence. At the more technical level, the issue of statistical inference has been discussed by philosophers of science and scientists working in parallel and sometimes in tandem. My concern here, however, is with the structure of scientific explanations. Whereas I believe that the problems of verification are basically the same in all disciplines, I shall argue that the differences in their subject matters impose different strategies of explanation. I shall distinguish between causal, functional and intentional explanations, corresponding – broadly speaking – to the physical, biological and social sciences respectively. While recognizing that causal explanation in some sense or senses is more basic than the other modes, I shall argue that there nevertheless is room and need for the latter.

The dichotomy between rational-choice theories of technical change and evolutionary theories corresponds – and once again I have to speak very broadly – to the distinction between intentional and functional explanation. It will turn out, however, that in many respects the exceptions to this statement are the more interesting cases. In fact, not only science, but philosophy of science as well, has most to offer at the level of detailed analysis where such glib generalities break down. Technical evolution differs from biological evolution in that the changes are far from totally random, but to some extent directed; they are also screened by a mechanism in which human intentionality plays a crucial role. Similarly, not all intentional models of technical change qualify as rational models – for there may be cases when the underlying expectations are not rationally formed. The lack of rationality in the expectations may be due to complex strategic interactions, or to a fundamental uncertainty about the future, or to both. It is perhaps the interaction between these two sources of ignorance that lends a unique flavour and depth to the issue of explaining technical change. Taken separately, both games without a solution and decisions under radical uncertainty create havoc with rational-choice models. When they both operate in a given choice situation, the result is close to chaos. Out of this chaos the evolutionary theories emerge as the more likely to explain actual technical progress. But here I go beyond my self-imposed limitations, so instead let me turn to matters more within my competence.

PART I
Modes of Scientific Explanation

Introduction to part I

The philosophy of science, generally speaking, has two main tasks. One is to explain the features that are common to all the sciences (or at least all the empirical sciences), the other to explain what sets them apart from each other. To begin with the second task, there is a long tradition of distinguishing between the natural sciences and the humanities *(Geisteswissenschaften)*. Within the natural sciences one may distinguish, furthermore, between the study of inorganic nature (or *physics)* and the study of organic nature *(biology)*. Within the humanities as traditionally defined there has developed a cleavage between the *social sciences* (which I define so broadly as to include linguistics, history, and psychology, in addition to the more obvious disciplines) and the aesthetic disciplines or *arts*. Now these distinctions by subject matter are not in themselves very interesting. Their relevance, if any, must stem from their being correlated with other classifications. I shall discuss three such ways of classifying the sciences: according to *method,* according to the underlying *interest,* and according to mode of *explanation.*

A widely held view is that the sciences are to be distinguished from each other according to their characteristic methods. The natural sciences, on this view, employ the hypothetico-deductive method, the arts use the hermeneutic method, and the social sciences the dialectical method. It is not always clear whether these are methods for theory construction or for theory verification, except that the hypothetico-deductive method clearly is of the latter kind. Let me briefly and without much argument state my opinions of this view. (i) The hypothetico-deductive method is *the* method for verification in all empirical sciences. If the hermeneutic method is understood as a procedure for verification, it can only be a sub-species of the hypothetico-deductive method. To be precise, the hermeneutic method is the hypothetico-deductive method applied to intentional phenomena, with some peculiar features due to the nature of these phenomena.[1] (ii) If the hermeneutic method is seen as a method for theory formation, it coincides with the notion of intentional explanation. (iii) The dialectical method as a procedure for verification invokes some kind of appeal to 'praxis', i.e. to the idea that social theories can simultaneously be agents of change and explanations of change. This, however, is ambiguous in that it can mean either that the theories are self-fulfilling or that they are instrumental in bringing about some desired change. I believe that the notion vaguely underlying most uses of the phrase 'the unity of theory and praxis' is that the theory should be both self-fulfilling and useful, but this, unfortunately, is normally not possible.[2] (iv) The dialectical method as a tool for theory formation can also be understood in several ways, the most interesting of which involves the notion of psychological and social *contradictions*. These, however, can be made intelligible in the standard causal-cum-intentional language of the social sciences.[3]

The upshot of this – excessively condensed – discussion is that there are no grounds for distinguishing between scientific disciplines according to their methods of verification, with the exception mentioned in note 1. Nor should hermeneutics or dialectics be thought of as methods for theory-formation that are somehow *sui generis*. In my view, there is equally little substance in Jürgen Habermas's theory that the sciences differ mainly in the interests they serve.[4] By his account, the natural sciences serve a technical interest, the hermeneutic sciences a practical interest, and the social sciences an emancipatory interest. Now this may be tautologically true, contingently true or contingently false, according to how the terms are further defined. As far as I can understand, the most reasonable reading of the view makes it come out as false. Each of the three scientific disciplines can serve each of the three interests, although perhaps to different degrees and (above all) in different ways. I do not want to enter into further discussion of this issue, since I believe that by any reading the theory is singularly unhelpful for the practising scientist – and this means that it fails the acid test for any philosophy of science. The language of interests is simply too coarse-grained and too external to scientific practice to mesh well with the fine grain of actual research.

I now proceed to sketch my own account of how the sciences differ from each other. I shall argue that the most illuminating and fertile distinction is between various modes of scientific explanation, which again are closely linked to strategies of theory-formation. Only certain kinds of theories are likely to have explanatory success in a given domain. On the one hand I distinguish between three modes of explanation: the causal, the functional, and the intentional. On the other hand, I distinguish between three domains of scientific research: physics (in the broad sense explained above), biology, and the social sciences. For reasons that are irrelevant to the present study I do not believe that the aesthetic disciplines can achieve or should aim at scientific explanation. Disagreement over this issue should not make any difference to the evaluation of the following arguments.

We may now ask: what kinds of explanation are adequate, proper and relevant for what domains of research? In Table 1 the two trichotomies are crossed with each other, so that whenever a given mode of explanation is clearly relevant for a given domain, the corresponding entry says 'Yes'; whenever it is clearly irrelevant, the entry is 'No'; and the cases in which there is room for doubt are entered as '?'. The dubious cases will ultimately turn out to be similar to the clearly irrelevant ones, but a good deal of argument will be needed to state the sense in which – and the extent to which – this is true.

Table 1

	Physics	Biology	Social science
Causal explanation applicable?	1 Yes	4 sub-functional explanation Yes supra-functional explanation	7 sub-intentional explanation Yes supra-intentional explanation
Functional explanation applicable?	2 No	5 Yes	8 ?
Intentional explanation applicable?	3 No	6 ?	9 Yes

The remainder of Part I will discuss in some detail what goes into the various categories in the table. Here I shall only offer brief comments on each of the nine entries.

(1) Causal explanation is the unique mode of explanation in physics, and physics is the standard instance and model of a science using causal explanation. In the brief remarks on physics in Ch. 1 below I shall mostly restrict myself to classical, pre-relativistic and pre-quantum-theoretical physics.

(2) Functional explanation involves, to anticipate, such notions as benefits, adaptation, selection and evolution. It might seem obvious that there is no room for these in physics, but some have thought differently. C. S. Peirce believed that the actually existing universe is but one of the many possible configurations of atoms, selected – by something like an absorbing Markov chain – because of its stability.[5] But I do not believe any physicists hold such views today. Although it is sometimes argued that the fundamental constants of physics change over time, no one to my knowledge has suggested that the change could be seen as evolution or progress.

(3) It is also generally agreed today that there is no rooom for intentional or teleological explanation in physics, but the consensus is a fairly recent one. The issue is also more complex than that of functional explanation in physics. The long Aristotelian tradition in physics, finally broken by Galileo, has I believe no plausibility today.[6] But teleological reasoning in physics by no means disappeared with the scientific revolution in the seventeenth century. For a long time it was thought that the least-time principle and other variational formulations of the physical laws had a teleological foundation. Light, apparently, was capable of behaving intelligently, of taking indirect routes in order to arrive at the goal in the shortest possible time.[7] Similarly, other processes in nature appeared to minimize action, energy or work. Since every physical process can be described as the minimization or maximization of something,[8] it is easy to fall into the belief that there must be a subject – a minimizer or maximizer – guiding the process. At least this was a plausible notion as long as physicists assumed a divine intentionality at work in the physical universe.[9] With the secularization of physics it is no longer possible to conclude backwards from maximizing behaviour to maximizing intentions.[10] Nor is there any other way in which, say, the behaviour of a pendulum can be explained by the tendency to minimize

potential energy.[11] The forces acting on the pendulum explain both why it moves the way it does and why this tends to minimize potential energy; the latter effect has no explanatory power of its own.

(4) Biology, of course, crucially invokes causal explanation. Functional explanation in biology, in fact, is justified only because we believe in the truth of a certain causal theory, viz. the theory of evolution by natural selection. By sub-functional causality in biology I refer to the random errors or mutations that on the one hand make possible natural selection and evolution, and on the other hand are responsible for such phenomena as senescence and cancer. These errors have no function, contrary to the popular idea that the function of mutations is to permit and further the evolution of life on earth.[12] By supra-functional causality I mean the causal interaction of many individual organisms, the behaviour of each of which can be explained functionally. There are, for example, biological analogues of 'unintended consequences' of individual behaviour.

(5) Biology is the paradigm for functional explanation, much as physics is for causal and social science for intentional explanation. The structure of functional explanation in biology is spelled out in Ch. 2. Here I shall only state the conclusion of that discussion: a structural feature or a behavioural pattern of an organism is explained functionally when shown to be part of a local maximum for the organism with respect to reproductive capacity.

(6) Intentional explanation in biology is hardly advocated by anyone today, as it was by the pre-Darwinian biologists, who could not explain the adaptation of organisms except by postulating a divine creator. Yet I believe that some biologists and philosophers of science have been led into underestimating the differences between functional and intentional adaptation, so that they in fact reason as if the evolution of organisms were guided by some intentional agency. They impute, that is, to natural selection the capacity to simulate some of the more complex forms of intentional adaptation, such as the capacity to use indirect strategies. Against this I shall argue that such simulation, while not impossible, is far less characteristic of natural selection than the tendency to come to rest at local maxima attained by incremental adaptation.

(7) Causal explanation in the social sciences can be understood in a framework partly similar to the one used in discussing biological causality. First, all phenomena covered by intentional explanation can also be

explained causally, although the relation is rather different from that which obtains between functional and causal analysis. Secondly, even those who would not go along with this view should accept that there is some room for causal explanation in the analysis of human behaviour. Sub-intentional causality is involved in the mental operations that are not governed by will or intention, but – metaphorically speaking – take place 'behind the back' of the individual concerned. Supra-intentional causality refers to causal interaction between intentional actors. (In addition there is also the intentional interaction between intentional actors which is studied by game theory.)

(8) The most controversial of the statements defended in the present work is the denial of a role for functional explanation in the social sciences. I am not here referring to functional explanation in the sense given above, involving the purely biological benefit of increased reproductive capacity. Such biological reasoning has a place in the study of human behaviour, although a very limited one. Here, however, I am concerned with explanations of social phenomena in terms of non-biological benefits. I would not dream of denying that such beneficial consequences can explain their causes even when the benefits are of a non-biological kind, provided that some feedback mechanism is specified. What I do deny, however, is that beneficial non-biological consequences can explain their causes even in the absence of a demonstrated feedback mechanism. In biology the theory of natural selection creates a presumption that whatever benefits reproductive capacity can also be explained by these benefits. In the social sciences there is no such theory of comparable generality, and so the actual mechanism must be specified for each particular case.

(9) The basic building block in the social sciences, the elementary unit of explanation, is the individual action guided by some intention. I shall argue that intentional adaptation differs from functional in that the former can be directed to the distant future, whereas the latter is typically myopic and opportunistic. Intentional beings can employ such strategies as 'one step backward, two steps forward', which are realized only by accident in biological evolution. I also discuss at some length the relation between intentionality, rationality, and optimality, arguing that these are less closely related than is sometimes assumed. There are irrational intentions, or plans that involve logical contradictions; there are also cases where rational action leads to satisficing rather than to optimizing.

Such, then, are the bare bones of the account that will be fleshed out in later chapters. To conclude this Introduction I shall raise and partially answer two related questions that are basic to my whole enterprise. These concern, first, the relation between causal explanations and the other modes of analysis; and secondly, the relation between macro-explanations and micro-explanations.

The reader may have asked himself the following questions. Is it really plausible to view causal, functional and intentional explanations as somehow being on a par? Is it not rather the case that causal explanation is *the* way of scientific reasoning, and that other explanations are merely convenient abbreviations or temporary expedients – second-best accounts that can in principle be dispensed with? And in particular, does not a materialist conception of the world imply that the goal of all scientific endeavour should be to explain everything in terms of cause and effect?

I have already suggested that functional explanation is indeed less fundamental than causal explanation. We can use functional explanation in biology because we have a causal theory – the theory of evolution by natural selection – that licenses us to explain organic phenomena through consequences that are beneficial in terms of reproductive capacity. Even if in a given case we are unable to tell the full causal story, we may be able to advance an explanation in terms of these beneficial consequences. True, this may lead us astray, because a phenomenon may have beneficial consequences that cannot be invoked to explain it, but in biology this is a fairly rare occurrence.[13] And if to 'beneficial' we add 'optimal', the frequency of non-explanatory benefits is probably close to zero.[14] It remains true that the functional explanation does not give the full answer to the question, 'Why is the organism organized the way it is?' Some features will be left unexplained, and even the features that are explained, are so incompletely. We understand a good deal when we know that a feature is there because of its beneficial consequences, but we would understand it even better if we knew the whole story of how it came to be there. But to say that a glass is half empty, is to say that it is half full; functional explanation does explain, although incompletely.

The relation between intentional and causal explanation is far more intricate and controversial. To discuss it adequately is well outside my competence. I shall rest content, therefore, with setting out without much argument the view that to me appears the most plausible, that of

Donald Davidson.[15] On his view it is possible to be a materialist on the mind–body issue without also being a reductionist. Mental states just *are* brain states, but this does not mean that talk about mental states can be unambiguously reduced to talk about brain states.

To explain this a bit more, let me use 'x' as an individual variable and 'a' as an individual constant, denoting some specific person. Saying 'a believes that Beethoven died in Vienna' is a *statement* about the person a, while 'x believes that Beethoven died in Vienna' is a *predicate* which can be applied to a given person, such as a, by replacing 'x' with his name. Now Davidson is a materialist in that he believes that the fact expressed by the statement 'a believes that Beethoven died in Vienna' can be exhaustively described by some (presumably very complex) neuro-physiological statement. The statement, if true, is true because a certain person in a certain situation believes that Beethoven died in Vienna, and a full physiological description of the person in this situation would also contain a description (in physiological terms) of his belief that Beethoven died in Vienna. On the other hand, Davidson is an anti-reductionist in that he believes that the predicate 'x believes that Beethoven died in Vienna' is not extensionally equivalent to any neuro-physiological predicate. 'If a certain psychological concept applies to one event and not to another, there must be a difference describable in physical terms. But it does not follow that there is a single physically describable difference that distinguishes any two events that differ in a given psychological respect.'[16] To conclude from materialism to reductionism, then, is to commit the same fallacy as when from the fact that everything has a cause one concludes that there is something which is the cause of everything.

A consequence of Davidson's view is that actions are caused by beliefs and desires. Of course, to say that the agent's doing p was caused by the desire to do p, is not to give a causal explanation. It is only to rephrase the intentional explanation in such a way as to indicate the existence of some (unknown) causal explanation. Nevertheless I shall have occasion to use this language of beliefs and desires causing action, as it is very convenient for many purposes. In particular it is helpful in order to establish a coherent way of talking about irrational phenomena.

I believe, then, that intentional explanation is *sui generis* in a sense that does not hold for functional explanation. Using causal explanation, we can talk about all there is, including mental phenomena, but we shall not be able to single out mental phenomena from what else there is. An analogy from mathematics may prove instructive. There exists a decidable axiomatization of the real number system, in which, however, it is impossible to define the set of natural numbers. Indeed, this must be impossible, since otherwise we would have a decision procedure for the theory of natural numbers, which is known to be impossible.[17] Using the axiom system, we can talk about all the numbers there are, including the natural numbers, but we shall not be able to single out the natural numbers from what else there is.

Be this as it may, it is clear that for practical purposes we may treat intentional and causal explanations as wholly distinct. When using a functional explanation, we usually know at least roughly how it could be backed by a suitable causal story. But in the present state of knowledge, we cannot even see in outline how intentional explanations are related to causal analyses of the same phenomena. Even if at some time in the future Davidson should join the line of distinguished thinkers who have declared it impossible in principle to do something which scientists have then gone on to do in practice, philosophers of science can for the present safely take the two perspectives as being radically different.

My second question is about the kind of facts that can enter into a scientific explanation. Within a given scientific discipline there will usually be concepts defined at different levels of organization and complexity. Within physics there are molecules, atoms, elementary particles and so on in an apparently endlessly descending chain; within biology there are ecosystems, populations, organisms, organs, cells and genomes; within the social sciences there are societies, organizations, industries, firms, families and individuals. Problems and puzzles can be stated at any given level, but at which level should one seek the solution? In particular, can one explain facts at a highly aggregate level by other facts at the same level, or should one seek the explanation in lower-level facts?

Generally speaking, the scientific practice is to seek an explanation at a lower level than the explanandum. If we want to understand the pathology of the liver, we look to cellular biology for explanation. If we want to understand chemical bonding, quantum mechanics provides the explanation. If we want to understand social revolutions, we seek an explanation in individual actions and motivations. The *search for micro-foundations,* to use a fashionable term from recent controversies in economics,[18] is in reality a pervasive and omnipresent feature of science. It corresponds to William Blake's insistence that 'Art and science cannot exist but in minutely organized Particulars.' To explain is to provide a mechanism, to open up the black box and show the nuts and bolts, the cogs and wheels of the internal machinery. (Here the term 'mechanism' should be understood broadly, to cover intentional chains from a goal to an action as well as causal chains from an event to its effect.)

The role of mechanisms is twofold. First, they enable us to go from the larger to the smaller: from molecules to atoms, from societies to individuals. Secondly, and more fundamentally, they reduce the time lag between explanans and explanandum. A mechanism provides a continuous and contiguous chain of causal or intentional links; a black box is a gap in the chain. If we look at mechanisms from the first perspective, they are always relative. 'From the standpoint of either scientific investigation or philosophical analysis it can fairly be said that one man's mechanism is another man's black box. I mean by this that the mechanisms postulated and used by one generation are mechanisms that are to be explained and understood themselves in terms of more primitive mechanisms by the next generations.'[19] Viewed from the second perspective, however, it might appear that systems of simultaneous differential equations embody the case of *no* time lag between explanans and explanandum. If this were truly so, they would provide the canonical form of explanation in science, an explanation that could not be improved on. Such systems, however, are always artificial, since the variables are never disaggregated to the same extent. Relating macro-variables such as national income, investment and consumption by a set of differential equations may be convenient for analytical purposes, but clearly misrepresents reality. There must be lags between macro-variables.[20] We may conclude that although the purpose of a mechanism is to reduce the time lag between cause and effect, the success of the reduction is constrained by the extent to which macro-variables are simultaneously replaced by micro-variables.

1. Causal explanation

The analysis of the causal relation and of causal explanation raises deep philosophical questions, which would of necessity fall outside the scope of the present exposition even were they within my competence. Happily, I can now refer the reader to a full and excellent discussion, *Hume and the Problem of Causation*, by Tom L. Beauchamp and Alexander Rosenberg. I shall limit myself, therefore, to some fairly broad comments on the general issue of causation and causal explanation, and then discuss in somewhat more specific terms causal explanation in the social and historical sciences.

a. Causal explanation: in general

The issue of the nature of causation must be sharply distinguished from that of causal explanation. The causal relation holds between *events*, by virtue of a regular conjunction between events of that kind. This is the Humean view of causation, expounded and defended by Beauchamp and Rosenberg. The explananda of causal explanations, by contrast, can be any fact: the fact that an event occurred or the fact that some state of affairs obtains. For the time being I shall limit myself to the causal explanation of events, and postpone the discussion of the explanation of states of affairs. Another difference is that causal explanation is mind-dependent, whereas the causal relation is not. In citing an event as the cause of another event, as we do in causal explanation, we can only refer to some features of the cause. But it is of course the complete event which acts as the cause; indeed there can be no such thing as an incomplete event. '*Whether* one event caused another is a separate issue from *why* one event caused another. The description of an event through its causally irrelevant features will pick out the same cause, even though it will not provide an explanation of why it had this particular effect.'[1] The causally relevant features are those cited in the causal *law* stating the regular conjunction under which the pair of events is subsumed. Thus true singular causal statements do not always provide a causal explanation. They provide an explanation only when the features cited in the statement are also cited in the law under which the events are subsumed. The statement, assumed to be true, that 'the shelving of a copy of Hegel's *Phenomenology* caused the bookshelf to break' fails to explain, because the causally relevant feature, viz. the weight of the book, is not cited.[2]

Causal explanation, then, works by subsuming events under causal laws. This view of explanation is related to, yet different from, Carl Hempel's 'covering-law' or 'deductive-nomological' model of explanation. On Hempel's view, the laws sustaining the explanation are universal lawlike statements from which, given a set of initial conditions, the explanandum can be logically derived.[3] This account is not an analysis of *causal* explanation, since it allows for such non-causal laws as Boyle's Law, in which the temperature, volume and pressure of a gas vary functionally and symmetrically in a way that causal sequences do not.[4] More important, Hempel's account is not an analysis of causal *explanation,* since many derivations of the specified kind do not explain at all. Two important cases are what, following David Lewis, we may call *epiphenomena* and *preemptive causation*.[5] There may be a universal, lawlike and true statement to the effect that A is always followed by B, and yet B may be a mere epiphenomenon if A and B are both effects of a common cause C. In other words, the problem of epiphenomena is identical to that of distinguishing true (or explanatory) from spurious correlation. Also, there may be a true law that ensures the production of A, given some initial conditions, and yet what actually produces A in a given case (where those conditions obtain) may be a totally different mechanism, which, as it were, preempts the mechanism underlying that law. In other terms, the problem of preemptive causation is identical to that of distinguishing between necessitation and explanation.[6] I shall return to both of these problems later, in my discussion of the relation between causal statements and counterfactual ones.

The causal relation is generally thought to obey the following three principles: determinism, locality, temporal asymmetry. I shall briefly discuss them in that order, with particular attention to their role in constraining causal explanation. Exceptionally, the causal relation itself may violate these principles; much more frequently, causal explanations are put forward that do not obey them.

Determinism is the postulate that any event has a cause: a determinate set of causal antecedents that are jointly sufficient and individually necessary for its occurrence. The denial of determinism may take several distinct forms. The best-known is the idea of statistically random events, implying the existence of a probability distribution over the range of possible outcomes. Objective randomness in this sense is not totally indeterminate, since the law of large numbers permits very accurate predictions when we deal with aggregates of events. Less familiar is the idea of objective indeterminacy, meaning that we cannot even assign probabilities to the various possible outcomes – not because we cannot ascertain these probabilities, but because they are not there to be ascertained. The distinction is similar to the one between risk and uncertainty (Appendix 1), with the difference that in the present case the random or indeterminate character of the process is assumed to have an objective basis rather than to be rooted in the cognitive deficiencies of the knowing subject. Moreover, this comparison also suggests the distinction between indeterminacy over the range of outcomes and indeterminacy within a fixed set of outcomes.[7] From the explanatory point of view, indeterminism with regularity (objective randomness or a restricted set of possible outcomes) is clearly more palatable than indeterminism without regularity (unrestricted objective indeterminacy).

Quantum mechanics is based on the idea of objective randomness at the level of elementary particle behaviour. In this domain explanation can only be statistical. It may be second-best according to the standard canons, but science itself tells us that no better explanation will be forthcoming. The situation is wholly different when we appeal to statistical explanation in other domains, such as statistical mechanics or the study of social mobility. In these cases we assume that the individual entities under study are governed by deterministic laws and that in principle a deterministic explanation could be constructed. There may, however, be overriding practical reasons for setting the explanatory standards lower and resting content with a statistical explanation. I return to this issue below. Here I only want to emphasize the radical difference between statistical explanation in quantum mechanics and in other disciplines. When we move from the physics of elementary particles and upward to the higher levels of variables, we move first from randomness to determinism and then, in some cases, from determinism to randomness. Statistical mechanics and quantum mechanics are both theories of random processes, but between them is the deterministic theory of the behaviour of individual molecules.[8]

Local causality means that a cause always acts on what is contiguous to it, in space and time. Action at a distance is impossible. If a cause is said to have an effect distant from it in time or space, we assume that there must be a continuous chain from cause to effect, with no unbridgeable gaps in it. In the history of science this postulate played an important role in the seventeenth-century discussion between Newtonians and Cartesians. The latter, rightly in many ways, regarded Newton's theory as obscurantist and reactionary, since it invoked an occult force that could act where it was not. But of course their alternative, even if unobjectionable on metaphysical grounds, did not have the superb predictive power of Newtonian mechanics, and so it went underground for two centuries until it was rehabilitated by the modern theory of gravitation.

In the case of Newton vs the Cartesians, what was involved was spatial action at a distance. There is also, however, a problem about temporal action at a distance, or remote direct causation. This notion is called *hysteresis* and is invoked in the study of some magnetic and elastic phenomena that apparently exhibit some form of memory. Leibniz, with his unrivalled lucidity, recognized that, if taken literally, the idea was just as unacceptable as the better-known spatial analogue. In this case he had to argue against fellow-Cartesians, who tried to save Descartes's theory of motion by postulating that the 'force' of a body at any given time depends on the length of the period during which it has accumulated. This, he argued, was just as absurd as if one were to claim that the current wealth of a person depends on the time he has used to accumulate it.[9]

To postulate hysteresis is to violate 'a widespread scientific dogma that all aspects of historical knowledge can be replaced ultimately and in principle by a sufficiently deep structural knowledge of the current state of the phenomena in question.'[10] If nevertheless many models, especially in the social sciences, exhibit hysteresis, it is because we have not yet acquired this structural knowledge. Both probabilistic models and models involving remote temporal action reflect our ignorance of the local deterministic causality assumed to be operating.[11] Causal explanation in such cases has features not believed to be present in the causal relations sustaining them.

In some cases we may close the ignorance gap by invoking alternatively randomness or hysteresis. Or rather, we reduce the randomness by adding hysteresis. This is the case, for instance, with the higher-order Markov chains, in which the probabilities of transition from one period to the next depend not only on the current state of the system, but also on earlier states.[12] An example is weather prediction, where knowledge of the weather today and yesterday is a better predictor than just knowledge about today's weather. First-order chains, by contrast, embody the 'Markovian assumption' that the transition probabilities depend on the current state only. It is incorrect to argue that this assumption is the 'probabilistic analog to the principle of scientific determinism'.[13] First-order chains retain local causality, but violate determinism – and these are distinct regulative principles.

It will be clear by now that the principle of local causality is related to what I referred to earlier as the need for mechanisms in scientific explanation. Moreover, the twin concepts of local causality in space and local causality in time are related to the two aspects of mechanisms which I referred to as the substitution of micro-variables for macro-variables and of short time lags for longer lags. Yet local causality is a feature of the world, i.e. of the causal relation as it exists independently of our minds, whereas the notion of a mechanism is mind-dependent, since it only makes sense in contrast to more coarse-grained explanations. It is the belief that the world is governed by local causality that compels us to search for mechanisms of ever-finer grain. Finally it should be clear that local causality shows us the way out of the difficulties met by Hempel's theory of explanation, since the issues of epiphenomena and preemptive causation do not arise when we postulate continuous and contiguous causal chains. It is also clear, however, that any actual mechanism may be too coarse-grained to avoid these difficulties. The difference between Hempel's view and the present view is that I have argued that the explanation must be abandoned if and when a finer specification of the mechanism shows that what we believed to be causation was in fact an epiphenomenon or preemptive causation.

Temporal asymmetry means that a cause must precede its effect; or at least not succeed it. The controversial part of this principle concerns the difference between the two formulations just given. Can an effect be simultaneous with its cause? I shall return to this question later. Here I want to observe that the principle of temporal asymmetry can be generalized from causal explanation to any kind of explanation: *the explanans cannot succeed the explanandum*. For intentional explanation this follows from the fact that we do not explain intentional behaviour by the actual consequences that follow from it, only by the intended consequences, which may not be realized or may even be unrealizable. It might seem that functional explanation violates the generalized principle, since in this case phenomena are explained by their actual consequences.[14] But the conclusion to be drawn is not that retroaction is possible, but that the explanandum must be an entity which persists over time, not a one-shot event. If we were to say, for example, that crime is to be explained by its beneficial effects for capitalist competition, we can only mean that crime at time t has beneficial consequences that uphold and sustain crime at some later time t'.

The three principles of causality are logically independent. I shall argue for this only with respect to the first two principles, since the third does not present any difficulty (or interest) in this regard. With determinism without local causality, we invoke the past in order to explain the present and predict the future. Let us assume, for example, that the following statement has been shown to be false: whenever two systems (of a given kind) are identical at time t, they are also identical at all later times. From this we may conclude the falsity of determinism, but we need not do so. We can retain determinism if the following statement is true: whenever two systems have been identical up to and including t, they will be identical at all later times. For the first statement to be true and the second false, there must be some real hysteresis in the system. It has been suggested, for instance, that the apparent indeterminism of quantum mechanics can be done away with by taking account of the past history of the particles.[15] But the standard solution is the converse one, to retain local causality and abandon determinism. This involves rejecting both of the above statements, while accepting the following: whenever two systems are identical at time t, they will have identical probability distributions for their future development.

The epistemological status of the three properties of the causal relation is a difficult matter indeed. Perhaps the best answer is the following. The principles are invariably found to obtain at the molar level, that of objects roughly comparable in size to human beings. This has led, by the kind of mechanism so well described by Hume, to a belief in their absolute necessity, for objects smaller as well as larger than those we can manipulate and observe. And philosophers have sought to understand the nature of, and the grounds for, this necessity. Science, however, has ultimately come to question and – in the case of determinism – to reject the principles on empirical grounds. And if they can be questioned on empirical grounds, they would appear to be empirical in nature – even if at some later time they come to be reaffirmed. This last step, however, is not obvious. If future scientific analysis leads to a rehabilitation of determinism, it might do so in a way that makes it appear as truly necessary – in one of the many senses of that term.

Be this as it may, however, the working scientist outside particle physics may probably go on invoking the principles with a fairly good conscience, not only at the molar level, but also for most micro- and macro-phenomena that are outside the reach of the senses. But his conscience can be no better than fairly good, because the objective randomness of quantum mechanics cannot be wholly confined to particle physics. It may spill over into biology as well as psychology: biology, since the random mutations may ultimately have some quantum-mechanical explanation, and psychology, since the phenomena of thought may depend on the random firing of neurons.

There remain two loose threads to the argument, which in fact can be used reciprocally to tie each other up. These are the issue of causal explanations of states of affairs and the issue of simultaneous causation. That there can be causal explanations of matters of fact is obvious. The fact that my typewriter is at present located 85 cm above the floor is a state of affairs that can be causally explained. But of course this state does not enter as a relatum in a causal relation, since only events can be causal relata. A causal explanation of a state of affairs trades on causal laws, but in a more complex way than causal explanations of events. My typewriter's present position can be explained as the equilibrium outcome of a very large number of interrelated causal mechanisms that specify what goes on in the fine grain of the typewriter as well as in that of the table on which it is at rest. It is tempting to say that the typewriter's current position is explained, and perhaps even caused, by the fact that it is currently at rest on the top of my table. Thus we might introduce the notion of simultaneous explanation, and perhaps of simultaneous causation. But I believe these are temptations to be resisted, or at least indulged in metaphorically only. An equilibrium may be modelled as timeless, but this is a superficial approximation only. Equilibria are sustained by micro-mechanisms operating in finely attuned and compensating ways. This holds for the thermal equilibrium of a gas as well as for a general economic equilibrium. We may want to say, as a shorthand expression, that one feature of the equilibrium is explained by another, and this mode of expression may be useful for some purposes. But strictly speaking this is to commit the fallacy of epiphenomena, since the macro-features of equilibria are joint effects of the micro-structure.

b. Causal explanation: in the social sciences

Since the central topic of this book is the explanation of technical change – an issue arising within the social sciences – I have found it useful and in fact necessary to spell out how some of the general principles discussed above apply to these disciplines. I shall take up the following issues. What is 'the role of the individual in history'? What is the importance of hysteresis in the social sciences? What is the relation between causality and counterfactual statements in history and social science? What is the role of simultaneous causation in the social sciences? What is the actual and proper role of statistical explanation in the social sciences? Also, in Ch. 3 below I turn to some other modes of causal explanation in the social sciences, related to the causal explanation of beliefs and preferences. Needless to say, the discussion will be exceedingly sketchy and, I suppose, somewhat idiosyncratic.

The role of the individual in history. It follows immediately from what was said in the Introduction to Part I that in one sense this is a pseudo-issue. History is made by individuals and must be explained in terms of individual action. In another sense, however, the issue is meaningful. We may ask whether the course of history can be deflected by the action of a single individual. This is sometimes referred to as a question of determinism, but it really should be conceptualized as one of stability. We may ask, that is, whether societies have the property of *homeorhesis*,[16] so that for any small deviation in their course they will later on resume the development they would have taken had that deviation not occurred. If societies are stable in this sense, and if any action by an individual counts as a 'small' contribution, then the individual has no proper role in history.

It is easy to see how this issue comes to be confused with that of determinism. We often want to achieve two tasks simultaneously, prediction and aggregation. If the social group in question is stable in the indicated sense, we may succeed in both, since the fine grain does not really perturb the prediction. This means that the group behaves as if it were deterministic in the aggregate. Of course, there are no general reasons to expect such stability as we have in developmental biology, where the notion of homeorhesis was first put forward, and so we have to give specific reasons in each particular case to believe in dynamically stable societies. But assuming that some such demonstration has been given, we may uniquely predict the future value of macro-variables, using only present values of macro-variables. (Or possibly also past values, as I am here concerned only with determinism, not also with local causality.) If in other cases – which in the social and historical sciences will be an overwhelming majority – we are unable to predict macro-variables using only macro-variables, this does not mean that the society in question does not behave deterministically, only that it lacks a certain kind of stability. It may still be possible, however, to formulate macro-theories, if we renounce unique prediction. Attempts have been made, for example, to develop a theory of historical change by using *correspondences* (improperly called many-valued functions) rather than *functions* (redundantly called single-valued functions).[17] Given the current macro-state, we may be able to predict that the next macro-state will be found within some subset of the abstractly conceivable states. Or, alternatively, we might try to define a probability distribution over the set of macro-states that may come about.

Hysteresis in the social sciences.[18] It has been argued that hysteresis is of particular importance in the social sciences, because of the irreducible importance of history for the understanding of the present.[19] At the ontological level, this cannot be the case, as argued above. Yet it is probably true that in the social sciences models and theories exhibiting time lags have a more prominent place than in the physical sciences, because the structural knowledge that would enable us to do away with the apparent hysteresis has not been attained. The goal of research should be to substitute for past causes the traces left in the present by the operation of those causes, but this we are not always able to achieve. Such traces are usually called 'state variables', and they occupy a privileged position in scientific explanation.[20]

An example from the study of technology will prove useful later on. The neoclassical production function exhibits output as a function of aggregate capital and labour: $P = f(K,L)$. The so-called 'capital controversy' has shown, at least to my satisfaction, that the notion of aggregate capital in some cases defies construction.[21] If the underlying disaggregated technological structure has certain (not implausible) properties, the notion of aggregate capital does not behave as it is supposed to. In particular, it may not exhibit decreasing marginal productivity, nor verify the expected monotone relationship between the wage rate and the intensity of capital. Given this result, some authors[22] have argued that one should rather see the stream of output over time as a functional of the stream of labour inputs over time: $P(t) = F(L(t))$. This, however, involves hysteresis, because one cannot then predict output at time t without invoking labour inputs at earlier times. The procedure may be a convenient one for some analytical purposes, in addition to the obvious political appeal, but as a depiction of the causal structure of the process of production it is clearly inadequate. It must be possible to explain what goes on in the production process now without invoking the distant past. One way of doing this would simply be to disaggregate the production function: $P = f(K_1 \ldots K_n, L)$.

Causality and counterfactuals. It is usually taken for granted that there is a close link between causal statements and counterfactual statements, e.g. between 'A caused B' and 'If A had not occurred, B would not have occurred'.[23] Yet it can be shown that the truth of the counterfactual statement is neither sufficient nor necessary for the truth of the causal statement. The non-sufficiency is brought out by the possibility of epiphenomena. If C is sufficient for A and necessary for B, the counterfactual statement is true, yet the causal one false. The non-necessity is brought out by the possibility of preemptive causation. If C would have brought about B had A been absent, the causal statement remains true, but the counterfactual one is false. More generally, any attempt to define causal notions in terms of counterfacual statements involves putting the cart before the horse. I shall now argue that this accounts for the failure of two recent attempts to define the essentially causal notions of *power* and *exploitation*.

Power involves getting what one wants, but this alone will not do. One has to add (i) a causal clause to the effect that the power-holder was instrumental in bringing it about that he got what he wanted, (ii) that he intended to bring it about, (iii) that he brought it about in the way he wanted (i.e. not by a fluke), and (iv) that he brought it about in the face of resistance from other people, or, if there was no resistance, that he would have brought it about even had there been. The last clause shows that the analysis of power does indeed involve counterfactuals, but it would be a mistake to understand the causal component of power in terms of counterfactuals. If one argues that in the absence of the action of the power-holder, the effect would have been absent, one immediately comes up against the problem of preemptive causation. And if, following Alvin Goldman, one argues that the powerful man would get what he wanted even had he happened to want differently, one goes beyond what is implied by the notion.[24] To see this, consider the man who wants to bring about p because he knows that if he does not do so, his rival will, and so he preempts his rival out of spite, overcoming the rival's resistance. Surely this man has power to bring about p, even though his desire to bring it about is caused by the knowledge that it will be brought about in any case. And his power to bring about p is not, *pace* Goldman, lessened one iota by his inability to bring about q, should the desire to spite his rival take that form.

Exploitation involves taking unfair advantage of someone.[25] It thus invokes the moral notion of unfairness, as well as an essentially causal requirement that exploiter and exploited interact by the former 'taking advantage of' the latter. In his important work on exploitation and class, John Roemer has offered two definitions of exploitation: one which respects these constraints, and one which does not.[26] The latter – my only concern here – stipulates that a group is exploited if it would have been better off had it withdrawn from the larger society, taking with it some of the social assets according to some withdrawal rule. To overcome objections, he also stipulates that the complementary group – the exploiters – would have been worse off were they to withdraw under the same rules. But it is easy to construct examples in which two groups interact in a way that makes both of these counterfactual statements true, and yet the interaction is not one of exploitation.[27]

But although counterfactual statements do not exhaust the meaning of causal statements, they can offer true non-definitional characterizations of causal relations in many central cases. One reason why causality is so important practically is of course that frequently the effect will *not* be produced in the absence of the cause. Similarly, Roemer's definition of exploitation and Goldman's definition of power offer crucial insights into many central empirical cases.

Although the meaning of causality cannot be rendered by counterfactual statements, the latter have an important role in causal analysis. When we want to assess the relative importance of causes $C_1 \ldots C_n$ for some effect E, we have to remove in thought the causes, one at a time, to see what difference the absence of that cause makes for the effect. (And, of course, we must also remove any other cause that might be waiting in the wings but was preempted by the cause in question. I will return to this later.) When the effect E is some quantitative variable, the relative importance of the causes can be assessed simply by evaluating their quantitative impact on E. In other cases this may not be feasible, and then we must use our ingenuity to construct other measures. If, say, the absence of cause C_1 would have led to a delay of one year in the production of E and the absence of C_2 to a delay of 10 years, we are certainly justified in saying that C_2 was more important than C_1, but perhaps not in saying that it was 10 times as important. This procedure could be used, for example, in an analysis of the relative importance of the causes of World War I, while the more direct quantitative approach could be used to evaluate the causes of American economic growth in the nineteenth century.

The difficult epistemological question is how to ascertain the truth, or the assertability, of counterfactual statements. In my view, it is hard to accept that such statements have truth conditions, i.e. that they are true, when they are true, because of some correspondence with the facts. In particular, I cannot accept the suggestion that counterfactual statements should have their truth conditions in *possible worlds*, as suggested by several recent writers.[28] They argue that the statement 'If A had been the case, then B would have been the case' is true in world x if and only if (i) A is not the case in any world that is possible relatively to x, and (ii) B is the case in the world closest to x in which A is the case. Here it is assumed, first, that the notion of a possible world is made relative to another world and not just seen as possible *tout court*, and, secondly, that we can somehow measure the distance between a given world and the other worlds that are possible relative to it.

My own view, which is also that of other Humean writers on causation, is that counterfactuals cannot be true or false, only assertible or non-assertible.[29] The grounds or conditions for assertability are our currently accepted scientific theories, i.e. the currently accepted lawlike universal statements. A counterfactual statement is warranted or assertible if the consequent can be deduced from the antecedent together with some suitably chosen theoretical statements. On the other hand, it is important to retain part of the possible-world account, viz. the idea that we should restrict ourselves to the *closest* alternatives under which the antecedent is true. We want to say that the counterfactual is assertible when the consequent follows on the assumption of the antecedent *and the minimal number of other changes*. When we assume a contrary-to-fact antecedent, we normally have to add a number of other changes to make the assumption internally consistent.[30] We cannot, say, remove in thought all slaves from the pre-1861 American South, without also removing slave overseers, slave owners, slave laws etc. But although some changes of this kind must be accepted, not all changes have to be. If we want to defend the statement 'In the absence of slavery the Southern economy before 1861 would have had a substantially higher rate of growth', we are not committed to a high rate of growth in *all* internally consistent no-slave pre-1861 economies, including those devastated by earthquake or plagues.

In addition to the internal consistency of the counterfactual assumption we should require historical possibility. If the antecedent of an historical counterfactual cannot be inserted into the actual historical development, it should not be entertained. If we want to evaluate the actual American economy in 1860 by comparing it to a counterfactual state of the economy at the same time, we must require that the latter could have branched off from the actual history at some earlier time, for all that we know, i.e. for all that our theories tell us. For if this requirement is not fulfilled, we shall not be talking about the course of *the* American history, which is an historical individual.[31] Of course, for such branching-off to be possible our theories must have a non-deterministic character, but as argued above, this is only what we would expect at the level of the macro-theories used in the study of social change.

An example involving technical change may help to bring out what is implied by this approach to counterfactuals. Robert Fogel's study of the importance of railroads for American economic growth in the nineteenth century is probably the most ambitious attempt at counterfactual history ever made.[32] He sets out to defend a thesis: that the American GNP in 1890 would not have been much smaller had the railroad never been introduced.[33] In marshalling the arguments for the thesis, Fogel generally takes great care to follow the procedure outlined above: to reconstruct the American economic development from the latest possible time at which a non-railroad economy could have branched off, i.e. from approximately 1830. But there is one conspicuous flaw in the reasoning, pointed out by Paul David in his review of Fogel's book.[34] This is Fogel's implicit assumption that *introduction gains equal withdrawal losses*. David's main example concerns the existence of irreversible economies of scale, due to learning by doing (Ch. 6). If the smaller transportation costs of the railroad led to lower prices of – and greater demand for – transported goods, mass-production techniques may have developed that – assuming an instantaneous withdrawal of the railroad in 1890 – would still have larger efficiency at the *smaller* volume of production than the initially given techniques. Fogel, when measuring the lack of introduction gains by the loss from withdrawal, argues as if it were possible to measure the importance of the ladder at the point where it has become possible to kick it away.

The analysis of counterfactual statements in history gives rise to several puzzles, of which I shall mention three. The first we may call the 'scissors problem' created by the twofold use of theory in the analysis of counterfactuals. Clearly we need theory to go from the antecedent to the consequent in counterfactual statements, but we also require theory to evaluate the legitimacy of the antecedent taken by itself. We need theory to tell us whether the assumption of the counterfactual antecedent is compatible with what is retained in the actual world. Economic theory tells us, for instance, that we cannot normally assume a higher volume of sales and keep the unit price constant.[35] The puzzle arises because these two roles assigned to theory may come into conflict with each other. The better our theories, the more antecedents will be ruled out as illegitimate – and the more able we are to assess the consequent. Or to put it the other way around: the more questions we can ask, the fewer we can answer. Clearly there is a delicate balance to be struck here.

The railroad example, once again, will be useful. If we want to know what would have happened had the railroad not been invented, we might be interested in determining whether the internal combustion engine would have been invented before it actually was. To answer this question, we should need a theory relating technical change to various socioeconomic and scientific conditions. But if we had such a theory, the chances are good that it would also tell us that the invention of the railroad was inevitable, given the conditions existing at the time, so that the very theory telling us how to answer the question (what would have happened without the railroad?) also tells us that this is not a question we should ask (because the invention was inevitable according to the theory). If we assume away the railroad in 1830, the theory tells us that it will be invented in 1831; and if we insist on assuming it away throughout the period, theory tells us that something fairly close to it will be invented.[36] Preemptive causation, in other words, raises its ugly head again. Fogel is able to finesse this problem because he is not out to assess the exact importance of the railroads for American economic growth, but to prove the thesis that the importance was below a certain upper bound. He can then play the Devil's Advocate and assume that no substitutes for the railroad would have developed, for if his thesis is valid on this assumption, it will remain so when it is dropped.

The second puzzle arises when causes interact non-additively. Assume, schematically, that the causes C_1 and C_2 contribute to the effect E in the following way: $E = 2C_1 + 3C_2 + C_1 \cdot C_2$. Assume, moreover, that in a given case C_1 and C_2 have values 2 and 1 respectively, implying a value of 9 for E. If we try to determine the relative importance of the causes by assuming away first C_1 and then C_2, we get the result that C_1 contributes 6 to E (since without it the effect would be 3) and C_2 contributes 5 (since without it the effect would be 4). But this in turn implies that the total contribution of the two causes to the effect exceed the total effect, which is absurd.

The problem is acutely relevant for the theory of income distribution. If one accepts that factor shares should reflect factor contributions ('To each according to his contribution'), then we need a principle for determining what portion of the final product can be attributed causally to the various factors of production. If we assume that the factors are capital and labour (neglecting the problem of aggregating capital), related as in a Cobb-Douglas production function $P = a \cdot K^b \cdot L^c$, with $b+c=1$, then the counterfactual method has the absurd implication that each factor is causally responsible for the whole product, since the output is zero when one factor is zero. The approach chosen in some more simple-minded versions of neoclassical theory is to assume instead that the contribution per unit of each factor should be equated with the marginal product of that factor. It then turns out, magically, that the total contributions of all units of both factors exactly exhaust the total product.[37] And since the contribution of each worker is seen as the marginal product, the principle 'To each according to his contribution' immediately implies that labour should be paid according to marginal productivity.

To this line of argument one can raise the following objections. First, there simply is no way of decomposing the total product into the contribution of capital and that of labour. It is clear, in fact, that one cannot treat each worker as if he were the last to be hired.[38] This causal analysis makes sense at the margin, but not for the factor contribution as a whole. If we drop the assumption that $b+c=1$ in the production function, permitting increasing or decreasing returns to scale, then it is no longer true that the product is exactly exhausted by total factor 'contributions' (in the sense of marginal products). With increasing returns to scale, the sum total of 'contributions' will exceed the total product, with decreasing returns it will fall short of it. Secondly, the principle 'To each according to his contribution' should not be extended to the contribution of capital. Even though capital goods have a causal role in the production process, their owners do no work that could justify their retaining a part of the net product. Marxists often express this by saying that capital 'really' is frozen labour, and so cannot have a claim on any part of the product.[39] I have argued above that this language is misleading, as it gives the impression of some real hysteresis in the production process. The representation $f(K,L)$ of the production process is misleading because of the aggregation problem, but it would be wrong to dismiss this representation by arguing that capital, being only frozen labour, should not figure in the production function. If the aggregation problem did not arise, one might accept $f(K,L)$ as an adequate representation and yet resist the right of capital to a share in the product.

The third puzzle has to do with the identification of the actors involved in the counterfactual statement. We often want to know whether the economic situation – misery or wealth – of a certain group of people can be imputed to the economic system in which they are located. We then ask, counterfactually, whether they would have been better off in some alternative system. Clearly, it may be conceptually difficult to determine which system is the relevant one for comparison.[40] A more basic problem, however, is that in the alternative system under discussion it may be impossible to identify the group whose situation we want to assess, if it is defined in terms specific to the actual system. In that case we can evaluate the situation of the group only in the very special cases that the best-off member in the actual society is worse off than the worst-off member in the alternative system, or the worst-off member in the actual society better off than the best-off in the alternative.

Slavery once again provides an instructive example. A historian of American slavery writes that 'had it not been for plantations and slavery, the cities and towns of the South would have been even fewer and smaller, resulting in even fewer opportunities for nonslaveholding whites'.[41] Now, disregarding the 'micro-economic illusion' underlying the argument,[42] it fails on the simple logical grounds that in a Southern economy without slaves it would not be possible to identify the groups that in the slave economy were 'nonslaveholding whites' and to distinguish them from the slaveholders. This is so because the group of nonslaveholding whites are defined in terms of the very system which we are asked to assume away. There would be no problem, by contrast, if we asked about the fate of the small farmers or of the blacks in a system without slavery, since these are not categories defined in terms of slavery. Such problems of 'trans-world identity' arise fairly often in the social sciences, and tend to undermine our intuitive notions of what comparisons between groups or systems can meaningfully be made.[43]

The three puzzles I have singled out for discussion are all, basically, problems of aggregation. They arise because we often want to explain at the macro-level, even though the underlying mechanisms are of a finer grain. Clearly, the social sciences cannot in the foreseeable future dispense with macro-explanations, but neither can they expect to get rid of such puzzles.

Simultaneous causation. In the social sciences one constantly comes across such statements as 'democracies are conducive to scepticism' or 'life in big cities favours crime', in which contemporaneous states are related to each other as cause and effect. To make sense of such statements, without invoking the mysterious notion of simultaneous or instantaneous causation, we must be able to distinguish between exogenous and endogenous variables in the system under consideration. We can then interpret such statements in terms of the steady-state effect on an endogenous variable produced by a change in an exogenous variable. If we just consider the system at a given time after the change in the exogenous variable and observe that it has brought about a change in an endogenous variable, we cannot conclude that this is a steady-state effect. In the first place, the system may not yet have found its steady state, so that later the endogenous variable may take on quite different values; and in the second place the system may be such that there is no steady state towards which it converges after the change in the exogenous variable.[44]

I shall illustrate these principles by some examples from the work of Alexis de Tocqueville. In *Democracy in America* and elsewhere, he was concerned to assess the social consequences of constitutions, and in particular of democratic constitutions. In the course of this analysis he shows himself to be exceptionally sophisticated in his understanding of social causality. I shall briefly set out the four basic principles underlying his discussion.

The first we may call *the principle of holism:* the causal relations that are valid at the margin cannot be generalized to the whole. For example, even if love marriages tend to be unhappy in societies where they form the exception, one cannot conclude that they will also be unhappy in democratic societies where they form the rule. To marry for love in an aristocratic society is to court disaster, since going against the current tends to create hostility and in turn bitterness. Moreover, only opinionated persons will go against the current in the first place – and being opinionated is not a character feature conducive to a happy marriage.[45]

The second principle is that of the *net effect*. A typical example is the following: 'As there is no precautionary organization in the United States, there are more fires than in Europe, but generally they are put out more speedily, because the neighbours never fail to come quickly to the danger spot.'[46] A similar argument is applied to the impact of democracy on social integration, where Tocqueville argues that, compared to aristocracies, each person is tied to a greater number of persons, although the strength of each tie is weaker.[47] The general structure of these arguments is the following. We want to understand the effect of the exogenous variable, democracy, on such endogenous variables as the number of houses destroyed by fire or the strength of social integration. Tocqueville then observes that the effect in each case is mediated by two intermediate variables; that they interact multiplicatively rather than additively; and that they work in opposite directions. The net effect, therefore, could go either way, in the absence of information about the relative strength of the two tendencies. The methodological point is that the impact of democracy can be decided only by looking at the net effect, rather than by focusing on one of the partial mechanisms, as one might easily do.

The third principle is that of the *long term*. It is in fact a special case of the second, but of sufficient importance to be singled out for separate consideration. Tocqueville writes that 'in the long run government by democracy should increase the real forces of a society, but it cannot immediately assemble, at one point and at a given time, forces as great as those at the disposal of an aristocratic government or an absolute monarchy.'[48] The point is similar to that made by Schumpeter in a famous argument for capitalism, further discussed in Ch. 5 below.

The fourth principle is that of the *steady state*. Perhaps the central argument of Tocqueville's work on democracy is that one should not confuse the effects of democratization with the effects of democracy. The former are the effects that are observed before democracy has found its 'assiette permanente et tranquille',[49] whereas the latter are found when the process finally comes to rest. Observe that the distinction between short-term and long-term effects is made *within* the steady state: the short-term inefficiency and the long-term efficiency of democracy both belong to its steady-state features. One should not confuse, therefore, transitory effects with short-term effects. Tocqueville is well aware of the fact that there may be *no* steady state following from a given exogenous change. Writing around 1855 about the impact of the French revolution, he notes that 'I have already heard it said four times in my lifetime that the new society, such as the Revolution made it, had finally found its natural and permanent state, and then the next events proved this to be mistaken'.[50] In such cases, it is not possible to identify the effects of democracy: one can only look for the chain of effects stemming from the democratic revolution. But in his analysis of the United States, Tocqueville clearly believed it possible to distinguish the steady-state effects from the merely transient ones.[51]

44 MODES OF SCIENTIFIC EXPLANATION

In addition to the two difficulties raised above – that the system we observe may not yet have reached its steady state, and that it may not have a steady state – there is the difficulty that there may never be time for the steady-state effect to work itself out, because the system is constantly being exposed to exogenous shocks. In modern rapidly changing societies this is probably often the case. (Cf. Ch. 3 for a discussion of an analogous problem which arises in the models of social change as an absorbing Markov chain.) For all these reasons, it may be difficult to identify cases of simultaneous causation. On the other hand, one may alleviate the problem that stems from the absence of a steady state by extending the notion so as also to include *limit cycles*. If, after a change in an exogenous variable, the endogenous variables ultimately settle down in a stable oscillatory pattern, this might be called an effect of the change in question. It has been argued, for example, that one effect of the French Revolution was to introduce a cyclical change between Orleanism and Bonapartism.[52]

Statistical explanation. Even though deterministic explanation is the ideal in science, one often has to rest content with less. In particular, statistical explanation may offer a partial understanding of the phenomena under study. I shall briefly discuss three varieties of this mode of explanation: deductive-statistical explanation, inductive-statistical explanation, and correlation analysis.

Following Hempel,[53] deductive-statistical explanation deduces a statistical regularity from the assumption that a *stochastic process* is operating. In the study of social mobility, one may try to explain the observed mobility patterns on the assumption that we are dealing with a first-order or a higher-order Markov process, for instance. We assume, that is, that at any given time the probability of transition from one group to another depends on present and past group membership. This approach satisfies the demand for a mechanism in the sense of showing how a simple micro-process generates complex macro-patterns, but it also leaves us unsatisfied because the process, though simple, remains fundamentally obscure as long as no theoretical justification for the model is provided. The statement, say, that a worker will remain a worker with probability 60 per cent and undergo upward mobility with probability 40 per cent, cannot be the explanatory bedrock. In fact, if there is no theoretical justification for the model it can hardly be said to explain at all, since the apparent ability to account for the data may just be a case of 'curve-fitting' – a notorious practice in the social sciences.

There are several ways in which one might attempt to provide a theoretical underpinning of the transition probabilities. (i) Assuming that the choice of career depends on preferences, ability and opportunities, the random character of the former may be made intelligible in terms of the random distribution of one of the latter. Ability, to take the most obvious case, may be understood in terms of statistical genetic laws. Although this is just passing the explanatory buck, it may improve our understanding if the theory invoked satisfies one of the presently discussed requirements. (ii) The choice of career may be understood through the idea of a mixed strategy, i.e. an intentionally chosen probability distribution over the set of possible strategies. What looks like an imperfect causal explanation is transformed into a fully satisfactory intentional one.[54] (iii) The context may be such as to justify an assumption of equiprobability, based on the principle of insufficient reason. To take an example outside the social sciences, in statistical mechanics one derives the Maxwell distribution of the velocities of the molecules from the assumption that their motion is random, i.e. has no bias in any particular direction. The assumption of randomness is justified by the following argument:

> If the gas consists of a very large number of moving particles, then the motion of the particles must be completely random or chaotic.... If the motion were orderly (let us say that all the particles in a rectangular box were moving in precisely parallel paths), such a condition could not persist. Any slight irregularity in the wall of the box would deflect some particle out of its path; collision of this deflected particle with another particle would deflect the second one, and so on. Clearly, the motion would soon be chaotic.[55]

The argument assumes that there is no external force such as gravity making for regularity, nor any internal mechanism making for homeostasis. In the derivation of the Maxwell distribution one also has to assume that the particles have the same mass. It will be clear why arguments of this kind rarely apply in the social sciences. People differ from each other; they are subject to common forces not generated by their interaction; and they are not always deflected in their path when they meet.

Inductive-statistical explanation (another term taken from Hempel)[56] invokes statistical laws to explain particular instances rather than patterns of instances. An example would be the explanation of recovery from an illness by treatment with some medicament. Given the fact that most patients with the illness recover from it by this treatment, particular instances of recovery may be explained. Or again, a particular case of upward social mobility may be explained by pointing to the high transition probabilities linking the two states in question. Clearly, such explanations share the difficulty noted in the case of deductive-statistical explanation, and remain obscure as long as the probabilities are not justified theoretically. In addition, inductive-statistical explanation is beset with a difficulty of its own, pointed out by Hempel.[57] Consider the following inferences:

(1a) If the barometer is falling, it almost certainly will rain
(2a) The barometer is falling
(3a) It almost certainly will rain

(1b) When there are red skies at night, it almost certainly won't rain
(2b) The sky is red tonight
(3b) It almost certainly won't rain

Here all four premises may be true, and yet both the conclusions cannot be true. There must be, therefore, something wrong with the mode of inference. Hempel diagnoses the fault in the fact that the conclusion is detached from its premises, and argues that it can only be understood relative to the evidence or support on which it is based. This difficulty carries over from prediction to explanation, since 'no matter whether we are informed that the event in question . . . did occur or that it did not occur, we can produce an explanation of the reported outcome in either case; and an explanation, moreover, whose premises are scientifically established statements that confer a high logical probability on the reported outcome'.[58] Or again, consider the following example. Jones recovers from an illness after treatment by penicillin. Since most people treated for that illness by penicillin do recover, we can explain Jones's recovery on inductive-statistical lines. It may be the case, however, that a small subgroup of people are highly likely to be immune to treatment by penicillin, Jones being one of them. When nevertheless he recovered, it was because he belonged to a subgroup of that subgroup, the members of which are highly likely to be helped by penicillin. Clearly, the explanation – if this term is at all appropriate – of Jones's recovery must invoke his membership of the subgroup of the subgroup, rather than his membership of the general population. But if one knew nothing about these finer distinctions, one might well conclude that Jones recovered because penicillin in general tends to induce recovery. Inductive-statistical explanation, for the reasons brought out in such cases, is liable to be spurious.

Correlation analysis, and its more sophisticated derivations such as path analysis, proceed by finding systematic co-variations among variables. For a given dependent variable we may find that it co-varies with several independent variables, and by well-known statistical techniques it is possible to determine the relative strength of these links, to distinguish between direct and indirect effects, etc. Typically the dependent variable is only partly explained by the independent variables, and the remaining 'unexplained residual' is often very large. To say that an explanation is only partially successful is not, however, to say that it is spurious.[59] By the latter one usually refers to a correlation between two variables that does not stem from a causal relation between them, but from their common relation to some third variable. This is what was referred to earlier as the problem of epiphenomena. The danger of confusing correlation and causation is a constant problem in this mode of statistical explanation, and the reason why one should beware of interpreting correlations as more than indications that 'there is something going on' worth looking at in more detail. Or perhaps the point should rather be stated negatively, that the function of correlation analysis is that it enables us to discard a causal hypothesis if the correlation turns out to be low.

It is usual to say in such cases that one may distinguish true (i.e. explanatory) from spurious correlation by 'controlling for' the third variable. If, say, there is a high negative correlation between x (the percentage of members of a group who are married) and y (the average number of pounds of candy consumed per month per member), we might suspect that this is due to their being effects of a common cause z (the average age of the group members). Keeping z constant might in fact make the correlation between x and y disappear, and we might want to conclude without more ado that the correlation was a spurious one. And in this case this would probably be the correct conclusion to draw. But consider the following case, drawn, as was the preceding one, from an important article by Herbert Simon.[60] We assume a high positive correlation between x (the percentage of female employees who are married) and y (the average number of absences per week per employee). However, when z (the amount of housework performed per week per employee) is held constant, the correlation between x and y disappears. Here, however, we would rather conclude that z is an intervening variable between x and y. In the first case the causal chain is x←z→y, in the second it is x→z→y. In the first case the correlation between x and y was in fact spurious, in the second it was not, although in both cases it disappears when z is held constant. What enables us to distinguish between the two cases is *a priori* assumptions about the causal *mechanisms* that are likely to operate; and no amount of mere manipulation or controlling of the variables can substitute for such assumptions.

2. Functional explanation

Functional explanation in biology is, historically and logically, the prime example of this mode of explanation. Historically, because contemporary functionalist social science to a large extent derives from the biological paradigm; and logically since evolutionary theory remains the only wholly successful instance of functional explanation. I shall first, therefore, state this biological paradigm and then go on to discuss some varieties of functional explanation in the social sciences.

a. Functional explanation: in biology

I shall first sketch a highly simplified account[1] of the theory of natural selection, which is the foundation of functional explanation in biology. By keeping strictly to first principles I hope to avoid being too patently wrong on specific biological matters outside my competence. As the conclusions themselves will be on the level of first principles, I hope that my simplifications can be justified.

Let us think of the organisms in a population as a machine which constantly receives inputs in the form of mutations. (The machine analogy may appear strained, but will prove useful later.) For simplicity I assume asexual reproduction, so that mutations are the only source of genetic novelty; alternatively we may disregard recombination by arguing that in the long run only mutations can disturb the biological equilibrium (in a constant environment). I assume, crucially, that the mutations are *random* and *small*. The stream of inputs is random, in the sense that there is no correlation between the functional requirements or needs of the organism and the probability of occurrence of a mutation satisfying these needs. By mutagenes it is possible to increase the probability of mutations generally and even of structurally specified subgroups of mutations, but it is never possible – this being *the central dogma* of molecular biology – to modify the probability of functionally defined subgroups of mutations. Comparing mutations to misprints, one may increase the probability of misprints by breaking the glasses of the typesetter, but there is no way of selectively increasing the probability of misprints occurring in the second edition of a book that will correct the factual errors of the first edition.

Mutations, moreover, are assumed to be small, typically amino acid substitutions resulting from the misprint of a single letter in the genetic language. There are no doubt mechanisms, such as gene duplication, that can produce macro-mutations. In the first place, however, the evolutionary importance of these is at present highly controversial; and, in the second place, such mutations, while large compared to amino acid substitutions, are small compared to the discontinuities found in human innovation. To anticipate the discussion in Ch. 5 below, no gene duplication could produce a change of the order of magnitude of the switch from the horse-drawn carriage to the 'horseless' one.[2] Or to quote Schumpeter, 'Add successively as many mail coaches as you please, you will never get a railway thereby.'[3] As a main contention in the present work is that there is a basic difference between the local optimization through small improvements and the global maximization that permits steps of any size, the precise definition of 'small' is not really important.

In the phrase of Jacques Monod, natural selection operates by chance and necessity.[4] While mutations are random, the selection process is deterministic, in the sense that the machine at any given moment has well-defined criteria for accepting or rejecting any given mutation. (This means that I shall not deal with genetic drift or non-Darwinian evolution.) The mutation is accepted if the first organism in which it occurs benefits in the form of higher reproductive capacity. Since the organism then leaves more descendants than other organisms in the population, the new allele will spread until it is universally present. (This means that I shall not deal with frequency-dependent selection and other sources – such as heterosis – of stable polymorphism, save for a brief discussion in Ch. 3.). Once a mutation is accepted, the criteria for accepting or rejecting further mutations will typically change, since the organism, being now in a different state, may be harmed or benefited from different inputs. The machine says Yes or No to each input according to criteria that change each time it says Yes. If the machine ever arrives at a state in which it says No to each of the (finitely many) possible inputs, we say that it has reached a *local maximum*. The population climbs along a fitness gradient until it reaches a point from which all further movement can be downward only, and there it comes to a halt.

This *locally maximizing machine* is incapable of certain kinds of behaviour that, by contrast, are indissociably linked to human adaptation and problem-solving. First, the machine is incapable of learning from past mistakes, since only success is carried over from the past. In evolution there is nothing corresponding to the 'useful failures' in engineering.[5] Secondly, the machine cannot employ the kind of indirect strategies epitomized in the phrase 'one step backward, two steps forward'. A prime example of such behaviour among humans is investment, e.g. in machinery: consuming less now in order to consume more in the future. Thirdly, the machine is incapable of waiting, i.e. of refusing favourable opportunities now in order to be able to exploit even more favourable ones later. The patent system can serve as an example of such behaviour among humans: 'by slowing down the diffusion of technical progress it ensures that there will be more progress to diffuse'.[6] And lastly, the machine could not precommit itself, by taking steps today to restrict the feasible set tomorrow.[7] I discuss some of these issues more extensively in the next chapter. Here I only want to stress what has variously been called the impatient, myopic or opportunistic character of natural selection – it has no memory of the past and no ability to act in terms of the future.

A further analysis of the machine must take account of the possibility of environmental change. If the environment changes, the criteria for saying Yes or No to mutations will in general also change. A mutation is not beneficial or harmful in itself, only with respect to a given genetic background (itself the outcome of earlier mutations) and a given environment. With a changing environment it may well be the case that even transient local maxima are never attained if the organism is unable to keep pace with its surroundings. The notion of an environment, however, is ambiguous. In the first place, environmental changes may be modifications of the climatic and geological environment, to the extent that they cause evolutionary change without being themselves affected by it. (The last proviso is needed to exclude endogenous climatic changes, such as the change in the atmosphere generated by the evolution of plants.) In the second place, some parts of the environments are themselves evolving organisms or affected by such evolution. If a population is constantly subject to exogenous environmental change, a lasting equilibrium can never be attained, but to the extent that the environment is itself made up of (or affected by) evolving organisms, it makes sense to ask whether a *general biological equilibrium* – a state in which all organisms have attained local maxima with respect to each other – can be realized. *A priori,* it is not obvious that this will always be the case: there could be 'evolutionary games' without an equilibrium.

There is no logical objection to the idea of a world in which the rate of change of the environment is so high, relative to the mutation rates, that most organisms most of the time are badly adapted to each other and to their inorganic environment. In the world we know, however, the infinitely subtle adaptations found in the structure and the behaviour of organisms have for millenia evoked the wonder and – with less justification – the admiration of naturalists. In many well-documented cases the natural solution to structural and functional problems is strikingly close to what would have been chosen by an engineer working on the same problem. In some cases animals and men *are* facing the same problems, so that the actual solutions can be compared. As shown by d'Arcy Wentworth Thompson in his classic treatise *On Growth and Form,* as well as by several recent authors,[8] these solutions are often strongly convergent. In recent ecological work[9] nature is seen as an economist rather than an engineer; in fact optimal budgeting, linear programming, profit maximization, cost minimization, and even game theory are getting to be as much part of evolutionary theory as of economics. Evolution has been strikingly successful in solving the problem of hitting a moving target, i.e. adapting to a changing environment. This can only be because of the relative speed of the two processes involved: the speed with which evolution moves towards the target must be very much higher than the speed with which the target moves away from evolution. (I should add, to prevent misunderstanding, that a *regularly* changing environment can for the purposes of evolutionary theory be treated as a constant one.)[10]

We are now in a position to set out the logical structure of functional explanation in biology. I am dealing with the ideal case, in which we can account for the explanandum as fully as possible within the functional framework. From what has been said above, it will be clear that, and why, this ideal is rarely if ever fully realized. We may say, then, that a structural or behavioural feature of an organism is functionally explained if it can be shown to be part of a *local individual maximum with respect to reproductive capacity*, in an environment of other organisms which have attained similar local maxima. If we can show, that is, that a small change in the feature in question will lead to reduced reproductive capacity for the organism, then we understand why the organism has that feature.

I have discussed two of the elements in this definition: the notion of a local maximum and that of a general biological equilibrium. I now want to say something about the other two ideas that enter into the definition. First, we must insist on the strictly individualistic character of functional explanation in biology. Natural evolution promotes the reproductive capacity of the individual organism, not that of the population, the species or the ecosystem. In fact, enhancing the reproductive capacity of the individual may reduce that of the population. To explain how this can be so, we need some conceptual machinery that will turn out to be useful in later chapters.

Assume that an organism(or a human individual) may behave in one of two different ways, egoistically (E) and altruistically (A). The origin of whatever behaviour is observed, need not concern us here: it may be deliberate choice, mutation etc. Assume moreover that the organism (restricting ourselves to this case for the present) is living in a population of other organisms that can also behave in one of these two ways. Because of interaction the outcome – in terms of reproductive capacity – for the organism will depend not only on its own behaviour, but also on that of the other organisms. For convenience, and without much loss of generality,[11] we may assume that the situation confronting the organism is as follows. There are just four possibilities: on the one hand the organism itself may follow either A or E, and on the other hand either everybody else may follow A or everybody else follow E. We write '(A,E)' for the case in which the organism adopts A, and all others adopt E; '(E, E)' for the case in which the organism and others all adopt E; and so on. Taking an example from schooling in fishes,[12] let 'E' stand for the tendency to seek towards the middle of the school and 'A' for the absence of any such tendency. It is then clear that a mutation to E will be favoured, since, other things being equal, it is always better to be in the middle of the school than at the outskirts, because the fish in the middle are less exposed to predators. If, however, everyone seeks to be in the middle (as they will, since the fish doing so are favoured by natural selection), the school as a whole becomes more compact and more exposed to predators. This means that, from the point of view of the reproductive capacity of the individual fish, the four alternatives can be ranked in the preference order of the Prisoners' Dilemma:

Prisoners' Dilemma Preferences:
1.(E,A).2.(A,A).3.(E,E).4.(A,E).

For future reference, the following features of this structure should be noted. (i) If we restrict ourselves to the cases in which all behave in the same way, universal altruism is preferred over universal egoism. (ii) Egoism is a *dominant strategy*, since whatever the other fishes do, the best option for the individual fish is to adopt E. (iii) This means that the outcome will be that everyone follows E. (iv) This means that the actual outcome is *worse for all* than another conceivable option, and in particular worse for all than the situation which obtained before the mutation to E. The actual outcome is then *Pareto-inferior* to the initial state. (v) The 'collectively optimal' outcome (A,A) is both *individually inaccessible* (there is no advantage in taking the first step towards it) and *individually unstable* (there is an advantage in taking the first step away from it). (vi) For the individual organism, the best possible situation is one of unilateral egoism ('free rider'), the worst one of unilateral altruism ('sucker').

Many of these observations will prove more relevant when translated into the intentional language that is proper for the analysis of human behaviour. But they show clearly enough that there is no mechanism by which natural selection tends to favour the survival of the species or the population. The population, in fact, may 'improve itself to extinction'. This individualistic bias of natural selection does not, however, exclude the evolution of altruistic behaviour, by such mechanisms as kin selection, reciprocal selection or – conceivably – group selection.[13] But the bias does require that all such explanations have a firm basis in selection pressures operating on individuals. The older notion that features bringing collective benefits can be explained simply by those benefits is abandoned in contemporary biology.[14]

The last feature of functional explanation to which I want to draw attention is that the maximand is *reproductive* adaptation, not adaptation to the environment as measured, say, in life span. It goes without saying that some ecological adaptation generally is an indispensable means to reproductive adaptation, for if you don't survive, you can't reproduce either. But the connection is only a general one: what is favoured by natural selection is not the maximal degree of ecological adaptation, but only the degree which is optimal for reproductive adaptation. Too much ecological adaptation may be harmful for reproductive adaptation, if only because the very process of bearing and rearing offspring creates an ecological risk to the organism by exposing it more to predators than at other times. To maximize ecological adaptation an organism would have to be sterile, with zero reproductive adaptation.

It is important to see that the two last-mentioned features of functional explanation create a very different theory from the popular image of evolution. Instead of the comforting picture of natural selection adapting the species to its environment, e.g. by preventing overgrazing or aggression, we get the bleak story of individual organisms out to maximize the number of offspring, come what may. Or, even bleaker, a story about individual genes out to maximize the copies of themselves, using the individual organisms as their vessels.[15]

b. Functional explanation: in the social sciences

My discussion of functional explanation in the social sciences will proceed in two steps. First I shall set out an argument against such explanation that I have developed elsewhere, and that I still believe to be basically valid.[16] Secondly, however, I shall explain some of the reasons that have made me see that the issue is more complex than I used to think. Before I enter into the detail of these arguments, I want to say a few words about the immense attraction that functional explanation seems to have for many social scientists, quite independently of the serious arguments that can be marshalled in its defence. The attraction stems, I believe, from the implicit assumption that all social and psychological phenomena must have a *meaning*, i.e. that there must be *some* sense, *some* perspective in which they are beneficial for someone or something; and that furthermore these beneficial effects are what explain the phenomena in question. This mode of thought is wholly foreign to the idea that there may be elements of sound and fury in social life, unintended and accidental consequences that have no meaning whatsoever. Even when the tale appears to be told by an idiot, it is assumed that there exists a code that, when found, will enable us to decipher the real meaning.

This attitude has two main roots in the history of ideas. The first is the theological tradition culminating in Leibniz's *Theodicy,* with the argument that all the apparent evils in the world have beneficial consequences for the larger pattern that justify and explain them. True, this is not the only possible form of the theodicy, for there is also the alternative tradition that explains evil as the inevitable by-product of the good rather than a necessary means to it.[17] The breaking of the eggs does not contribute anything to the taste of the omelette; it just cannot be helped. Moreover, the theodicy cannot serve as a deductive basis for the sociodicy, to use a term coined by Raymond Aron, only as an analogy. There is no reason why the best of all possible worlds should also include the best of all possible societies. Indeed, the whole point of the theodicy is that suboptimality in the part may be a condition for the optimality of the whole, and this may hold also if the part in question is the corner of the universe in which human history unfolds itself. These logical niceties notwithstanding, the legacy of the theological tradition to the social sciences was a strong presumption that private vices will turn out to be public benefits.

Secondly, the search for meaning derives from modern biology. Pre-Darwinian biology also found a pervasive meaning in organic phenomena, but this was a meaning bestowed by the divine creator and not one that could serve as an independent inspiration for sociology. Darwin, however, gave biological adaptation a solid foundation in causal analysis and thereby provided a subtitute for the theological tradition to the demolition of which he also contributed. Formerly, both sociodicy and biodicy derived directly from the theodicy, but now sociodicy could invoke an independent biodicy. Once again, the biodicy served not as a deductive basis (except for some forms of social Darwinism and more recently for sociobiology), but as an analogy. In forms sometimes crude and sometimes subtle, social scientists studied society as if the presumptions of adaptation and stability had the same validity as in the animal realm. In the cabinet of horrors of scientific thought the biological excesses of many social scientists around the turn of the century have a prominent place.[18] The situation is less disastrous today, but the biological paradigm retains an importance out of proportion with its merits.

We may distinguish between the strong and the weak programme of functionalist sociology. The strong programme can be summed up in

Malinowski's Principle: All social phenomena have beneficial consequences (intended or unintended, recognized or unrecognized) that explain them.

This principle can be harnessed to conservative as well as to radical ideologies: the former will explain social facts in terms of their contribution to social cohesion, the latter according to their contribution to oppression and class rule.[19]

This theory was ably criticized by Merton,[20] who suggested instead the weaker programme that may be expressed in

Merton's Principle: Whenever social phenomena have consequences that are beneficial, unintended and unrecognized, they can also be explained by these consequences.

To locate the fallacy in this principle, let me set out what would be a valid, if rarely instantiated, form of functional explanation, in order to show how Merton's Principle deviates from it. From such standard sources as Merton's *Social Theory and Social Structure* and Stinchcombe's *Constructing Social Theories,* we may extract the following account of what a valid functional explanation in sociology would look like:[21]

An institution or a behavioural pattern X is explained by its function Y for group Z if and only if:

(1) Y is an effect of X;
(2) Y is beneficial for Z;
(3) Y is unintended by the actors producing X;
(4) Y – or at least the causal relation between X and Y – is unrecognized by the actors in Z;
(5) Y maintains X by a causal feedback loop passing through Z.[22]

There are some cases in the social sciences that satisfy all these criteria. The best-known is the attempt by the Chicago school of economists to explain profit-maximizing behaviour as a result of the 'natural selection' of firms by the market.[23] The anomaly that motivated this attempt was the following. On the one hand, the observed external behaviour of firms, such as the choice of factor combinations and of output level, seems to indicate that they adopt a profit-maximizing stance, by adjusting optimally to the market situation. On the other hand, studies of the internal decision-making process of the firm did not find that it was guided by this objective; rather some rough-and-ready rules of thumb were typical. To bridge this gap between the output of the black box and its internal workings, one postulated that some firms just happen to use profit-maximizing rules of thumb and others not; that the former survive whereas the latter go extinct; that the profit-maximizing routines tend to spread in the population of firms, either by imitation or by takeovers. If we set X equal to a certain rule of thumb, Y to profit-maximizing, and Z to the set of firms, we have an example of successful functional analysis, in the sense that it has the right kind of explanatory structure. Note, however, that condition (4) is fulfilled only if the rules spread by takeovers, not if they spread by imitation.

Russell Hardin has persuaded me that this example is not as unique as I formerly thought.[24] He gives, among others, the following ingenious example: the growth of the American bureaucracy can be explained by its beneficial consequences for the incumbent Congressmen, since more bureaucracy means more bureaucratic problems for the voter, and more complaints to their Congressmen, who are then re-elected because they are better able than new candidates to provide this service, but this also means that Congressmen have less time for legislative work, which then devolves on the bureaucracy, which therefore grows. Similarly, Skinnerian reinforcement may provide a mechanism that can sustain functional explanations, although once again it is doubtful whether condition (4) is satisfied, for 'if a causal link is so subtle that its perception is beyond the [beneficiary's] cognitive powers, it can play no role in reinforcement'.[25] If we drop condition (4), we get the class of explanations that might be called *filter-explanations,* in which the beneficiary is able to perceive and reinforce (or adopt) the pattern benefiting him, although in the first place these benefits played no role in its emergence. These explanations, while empirically important,[26] cannot serve as examples of successful functional explanation as the term is used here.

My main concern, however, is not with the rarity or frequency of successful instances of the paradigm set out above. Rather I want to argue that many purported cases of functional explanation fail because the feedback loop of criterion (5) is postulated rather than demonstrated. Or perhaps 'postulated' is too strong, a better term being 'tacitly presupposed'. Functionalist sociologists argue *as if* (which is not to argue *that*) criterion (5) is automatically fulfilled whenever the other criteria are. Since the demonstration that a phenomenon has unintended, unperceived and beneficial consequences seems to bestow some kind of meaning on it, and since to bestow meaning is to explain, the sociologist tends to assume that his job is over when the first four criteria are shown to be satisfied. This, at any rate, is the only way in which I can explain the actual practice of functionalist sociology, some samples of which will now be given.

Consider first an argument by Lewis Coser to the effect that 'Conflict within and between bureaucratic structures provides the means for avoiding the ossification and ritualism which threatens their form of organization'.[27] The phrasing is characteristically ambiguous, but it is hard not to retain an impression that the prevention of ossification *explains* bureaucratic conflict. If no explanatory claims are made, why did not Coser write 'has the effect of reducing' instead of 'provides the means for avoiding'? The term 'means' strongly suggests the complementary notion of an 'end', with the implied idea that the means is there to serve the end. But of course no actor deploying the means or defining the end is postulated: we are dealing with an objective teleology, a process that has no subject, yet has a goal.

As my next example I shall take the following passage from the third volume of Marx's *Capital:*

> The circumstance that a man without fortune but possessing energy, solidity, stability and business acumen may become a capitalist in this manner . . . is greatly admired by apologists of the capitalist system. Although this circumstance continually brings an unwelcome number of new soldiers of fortune into the field and into competition with the already existing individual capitalists, it also reinforces the supremacy of capital itself, expands its base and enables it to recruit ever new forces for itself out of the substratum of society. In a similar way, the circumstance that the Catholic Church, in the Middle Ages, formed its hierarchy out of the best brains in the land, regardless of their estate, birth or fortune, was one of the principal means of consolidating ecclesiastical rule and suppressing the laity.[28]

Once again we note the tell-tale use of the word 'means', as well as the suggestion that 'capital' – not to be confused with the set of individual capitalists – has eyes that see and hands that move.[29] True, the text may be construed so as to understand Marx as merely describing a happy coincidence – but in the light of his Hegelian background and persistent inclination to functional explanation, I cannot take this suggestion very seriously.[30]

Later Marxists have continued in the tradition of objective teleology. It is standard procedure among Marxist social scientists to explain any given institution, policy or behaviour by, first, searching for the class whose objective interest they serve and, next, explaining them by these interests. Or, more often than not it is assumed that all social phenomena serve the interest of the capitalist class, and then the issue becomes one of finding a suitable sense in which this is true. This is made very easy by the multiple meanings of the notion of class interest, which is ambiguous with respect to the distinctions between the interest of the individual class member and the interest of the class as a whole; between short-term and long-term interest; between transitional and steady-state interest; between immediate and fundamental interest; between economic and political interest. But of course the mere fact that some class interest is in some sense served, does not provide an explanation. It is true, for instance, that internal cleavages in the working class serve the interest of the capitalist class, but from this we should not conclude that they occur because they have this effect.[31] To do so is to neglect Simmel's important distinction between *tertius gaudens* and *divide et impera:* one may benefit from the conflict between one's enemies, and yet not be instrumental in bringing about that conflict.[32]

Marxist social scientists tend to compound the general functionalist fallacy with another one, the assumption that long-term consequences can explain their causes even when there is no intentional action (or selection).[33] There are, for instance, certain Marxist theories of the state that (i) reject the instrumental conception of the state as a tool in the hand of the capitalist class, (ii) accept that the state often acts in a way that is detrimental to the short-term interest of the capitalist class, (iii) argue that it is in the long-term interest of that class to have a state that does not always and everywhere act in its short-term interest, and (iv) assert that this long-term interest explains the actions of the state, including those which are against the short-term interest.[34] Now in the first place, the notion of long-term interest is so elastic and ambiguous that it can be used to prove almost anything; and in the second place one cannot invoke the pattern 'one step backward, two steps forward' without also invoking the existence of an intentional agent. One cannot have it both ways: both invoke an objective teleology that does not require an intentional agent, and ascribe to this teleology a pattern that only makes sense for subjective intentionality.[35]

This concludes my argument – or should I say diatribe – against functional explanation of the more unreflective kind. It might seem unfair to include Merton among, and even as the main exemplar of, the adherents of this procedure. He does in fact leave it an open question whether his functional analyses are intended to *explain,* or perhaps are meant only as a paradigm for the study of unintended consequences in general. But since sympathetic readers have taken his aim to be one of explanation,[36] and since this is certainly the interpretation that has been most influential, I feel that my presentation is largely justified.

The arguments given above have rested on two tacit premises. First, a functional explanation can succeed only if there are reasons for believing in a feedback loop from the consequence to the phenomenon to be explained. Secondly, these reasons can only be the exhibition of a specific feedback mechanism in each particular case. The second premise is not needed in the case of functional explanation in biology, for here we have general knowledge – the theory of evolution through natural selection – that ensures the existence of some feedback mechanism, even though in a given case we may be unable to exhibit it. But there is no social-science analogue to the theory of evolution, and therefore social scientists are constrained to show, in each particular case, how the feedback operates. I now go on to discuss two defences of functional explanation that rest on the denial of respectively the one and the other of these premises. I begin with the attempt by Arthur Stinchcombe to show that there *is* a social-science analogue to the theory of natural selection, and then go on to discuss the more radical proposal of G. A. Cohen that for a successful functional explanation there is no need for *any* knowledge – general or specific – of a mechanism.

Stinchcombe suggests that we should look at social change as an absorbing Markov chain.[37] To explain this idea, I shall use an example from a passage quoted above, concerning the function of conflicts within and between bureaucracies. We assume that the bureaucratic system can be in one of two states: R (for rigid) and F (for flexible). A rigid

62 MODES OF SCIENTIFIC EXPLANATION

bureaucracy has a hierarchical structure that does not permit the expression of conflicts. This leads to the accumulation of tension in the organization, which will have difficulties in adapting to changing conditions and problems. A flexible structure, on the other hand, permits the day-to-day-enactment of conflict and ensures adaptation. We then ask the Markov question: given that the organization at time t is in one of the states R and F, what are the probabilities that at time $t+1$ it will be in or the other of the states?

Table 2

		State at time t	
		R	F
State at time t+1	R	$0<p<1$	0
	F	$1-p$	1

The central assumption embodied in Table II is that the state F is an absorbing one. Once the organization has entered this state, it never leaves it, since there is no accumulation of tension making for change. If the organization starts in the state of F, it will remain there. If it starts in the state of R, we know from the theory of Markov chains – and it is in fact obvious – that it will sooner or later end up in state F with a 100 per cent probability and remain there. This shows, the argument goes, that there is a *presumption for equilibrium* in social systems. Non-equilibrium states are not durable, and so are replaced by others, that may or may not be in equilibrium. So long as there is a non-zero probability that the non-equilibrium state will be replaced by an equilibrium, the latter will sooner or later be attained.

The Markov-chain theory of social evolution differs from the natural-selection theory of biological evolution in that there is no competition between coexisting solutions to the same functional problems. Rather there is a sequence of successive solutions that comes to a halt once a satisfactory arrangement has emerged. In the language used above: once the machine has said Yes to one input, it stops scanning further inputs (until a change in the environment again makes it necessary). Or again, the theory assumes that social evolution is based on *satisficing* rather than *maximizing*. I shall have more to say about satisficing in Ch. 3 and Ch. 6 below.

What is the explanatory power of this ambitious and interesting attempt to provide a general basis for functional explanation in the social sciences? In my view it is feeble, since it is open to the two objections that I shall now state.

First, the model fails as a basis for functional explanation, since it does not explain social phenomena in terms of their beneficial or stabilizing consequences. Rather, the explanatory burden is shifted to the absence of destabilizing consequences. One implication is that the model violates our intuitive notions about functional explanation, which surely must be related somehow to *functions,* not just to the lack of dysfunctions. On the Markov-chain approach, for instance, we might say that the position of a sleeping person can be explained as an absorbing state: we toss around in sleep until we find a position in which there are no pressures towards further change.[38] This may well be a valid explanation of the position, but it is not a functional explanation. This, in itself, is not a serious objection. In the more general case, however, a state of functional neutrality or indifference will often induce drift,[39] which over time may accumulate to produce great changes. Tradition is a case in point. Unlike traditionalism, tradition has a short and inaccurate memory: it involves doing approximately what your parents did, not doing exactly what people in your society have been doing from time immemorial.[40] Traditional behaviour, therefore, is often in a state of incessant if imperceptible change. To explain traditional behaviour, such as the rain dance of the Trobrianders,[41] by the absence of destabilizing consequences is, therefore, to be a victim of myopia, or to accept unthinkingly the local description of the practice as ancient and unchanging. For tradition to be unchanging, there must be forces acting on it to keep the behaviour constant.[42]

In many cases, however, the absence of a positive effect will count as the presence of a negative one. For instance, it is not always true that those who pay the bureaucrats will continue to support them even if they do not deliver at least some of the goods. Bureaucracies sometimes survive merely by doing no harm, but this will hardly do as a general statement. By concentrating on stable equilibria, I can go on to state my second objection to Stinchcombe's model.

The Markov chain model applies to specific institutions, not to whole societies. For any given institution, what counts as an equilibrium will depend on the current state of all other institutions, just as in the biological case. This means that as in the case of biological evolution, we are dealing with a moving target. In the social case, however, there are no reasons for believing that the speed of the process of adaptation much, if at all, exceeds that of the change of the criteria of adaptation. On the contrary, the very example considered above brings out very well that social changes do not have the slow and incremental character of biological evolution. Once the bureaucratic system of a society changes as radically as from R to F, or from F to R, all other subsystems may be thrown out of – or move away from – equilibrium, and there may not be the time to move to (or substantially towards) a new one.

True, this is an empirical matter that admits of degrees. In traditional peasant societies it may more nearly be the case that all subsystems are in a state of mutual adaptation to each other, though once again this could well be an illusion due to the observer's limited view (or the local ideology). Edmund Leach, in his study of the Kachin, argues for a very long cycle of social change in this traditional society, invoking an analogy with population cycles in ecology.[43] Be this as it may, it seems clear to me that in modern industrial societies too much is changing for equilibria to have the time to work themselves out in the way suggested by Stinchcombe.[44] The association between social anthropology and functional explanation, therefore, may not be accidental.

Cohen's defence of functional explanation rests on epistemological considerations, not on a substantive sociological theory.[45] He argues that while knowledge of a mechanism is a sufficient condition for a successful functional explanation, and the existence of a mechanism a necessary condition, the knowledge is not a necessary condition. A functional explanation must be backed by something beyond the mere observation that the explanandum has beneficial consequences, but we need not invoke a specific mechanism for the backing. An alternative backing is provided by *consequence laws*, e.g. general lawlike statements to the effect that whenever some institution or behaviour would have beneficial effects, it is in fact observed.

This idea may be spelled out by means of an example. In his analysis of the variety and number of organizations in the United States, Tocqueville offers an explanation that, in paraphrased quotation, goes as follows. In democracies the citizens do not differ much from each other, and are constantly thrown together in a vast mass. There arise, therefore, a number of artificial and arbitrary classifications by means of which each seeks to set himself apart, fearing that otherwise he would be absorbed by the multitude.[46] Similarly, without referring to Tocqueville, Paul Veyne offers the same analysis of the Roman colleges, whose 'latent function' was to permit festivities in a group that was small enough for the intimacy required, while the manifest function was some arbitrarily chosen and irrelevant goal.[47] Now it is hard to see how these unintended, unrecognized, and beneficial effects of the groupings can also explain them, and in any case neither author offers a mechanism that could provide an explanation. Cohen, however, would argue that the analysis might yet be valid. Consider the following proposition that sums up Veyne's discussion:

(I) The Roman colleges can be explained by their beneficial effects on the participants.

It will not be contested by anyone, I believe, that we cannot uphold this proposition simply by pointing out the existence of the benefits. A possible backing could be the following:

(II) The Roman colleges can be explained by their beneficial effects on the participants, through the feedback mechanism X.

While accepting that this justification is as good as any, and indeed better than any, Cohen argues that we may also invoke a backing of the following form:

(III) By virtue of the consequence law Y, the Roman colleges can be explained by their beneficial effects on the participants.

This law, in turn, might have the following form:

(Y) In every case where the emergence of colleges (or similar associations) would have beneficial effects on the participants, such associations in fact emerge.

The structure of this law, as of all consequence laws, is 'If (if A, then B), then A'.[48] Such a law being established, we can, if in a given case we observe A occurring and leading to B, invoke the law and say that A is *explained* by its consequence B. The explanation is not in any way invalidated by our lack of knowledge of the underlying mechanism, though we do of course assume that there is such a mechanism by virtue of which the explanation, if valid, is valid. Thus we may proceed as in the standard case: (i) propose a hypothesis; (ii) seek to verify it in as many specific cases as possible, (iii) actively seek to falsify it by counterinstances; (iv) after successful confirmation and unsuccessful disconfirmation accord it the (provisional) status of a law; and (v) use the law to explain further cases. The explanations suggested by Tocqueville and Veyne are, therefore, more convincing taken together than they are in isolation, since each case may enter into the law explaining the other.

Cohen argues for this thesis by citing the state of biology before Darwin. At that time biologists had sufficient knowledge to be justified in formulating a consequence law to the effect, roughly speaking, that the needs of the organism tend to be satisfied, even though they had not proposed a correct account of the underlying mechanism. Lamarck, for instance, was quite justified and indeed correct in explaining the structure of organisms through their useful consequences, even if wrong in his sketch of a mechanism. Similarly Cohen argues that Marxism today is in a pre-Darwinian situation. We do have the knowledge required to explain, say, the productive relations in a society in terms of their consequences for the productive forces, even though we are as yet unable to provide a detailed analysis of the feedback operating (see Ch. 7 and Appendix 2 below).

My first objection to this account is that it makes it impossible to distinguish between explanatory and non-explanatory correlations. Whenever we have established a consequence law 'If (if A, then B), then A', this may express some underlying relationship that does in fact provide an explanation of A in terms of its consequence B, or in terms of its propensity to bring about that consequence.[49] There is, however, always the possibility that there is a third factor C, which explains both the presence of A and its tendency to generate B. In fact, I believe this to be true for Lamarck's account of biological adaptation. He did not establish an explanatory correlation, for he wrongly thought that ecological rather than reproductive adaptation was the crucial fact about organisms. Darwin showed that reproductive adaptation is the 'third factor', which explains both the features of the organism and their tendency to be ecologically adaptive.

An example from the social sciences is more to the point. In his analysis of authority relations in Classical Antiquity, Paul Veyne invokes Festinger's theory of cognitive dissonance in order to explain why the subjects so readily accepted their submission.[50] Not to accept the natural superiority of the rulers would have created an acutely unpleasant state – 'cognitive dissonance' – in the subjects, who, therefore, resigned themselves to a state of submission from which in any case they could not escape. The resignation, to be sure, was useful to the rulers, though far from indispensable. If necessary, they could have upheld – and when necessary did uphold – their rule by force. The fact remains, however, that it was useful; moreover it took place in precisely those circumstances where it would be useful, viz. those characterized by severe social inequality. A moderate degree of inequality would not induce resignation – but neither would it lead to revolt. We may formulate, then, a consequence law to the effect that 'Whenever resignation would be useful for the rulers, resignation occurs', and yet we are not entitled to explain the resignation by these consequences, since there is a 'third factor' – severe social inequality – which, explains both the tendency to resignation and the benefits to the rulers. We must conclude that what explains the resignation of the subjects to the status quo is that it had benefits for them – the benefits to the rulers being merely incidental. Cohen's account of functional explanation in terms of consequence laws falls victim to the problem of epiphenomena. To clinch the argument, we may invoke Herbert Simon's argument, cited in the last paragraph of Ch. 1 above, that it is impossible in principle to distinguish between real and spurious correlations without some *a priori* assumptions about the causal mechanism at work.

Similarly, the problem of preemptive explanation undermines Cohen's account. There could well be a non-spurious consequence law of the form given above, and yet the presence of A in some specific instance might be due to a quite different mechanism, which preempted the mechanism underlying the consequence law. One could modify the example used in the previous paragraph to illustrate this possibility. Imagine that in general the rulers have to indoctrinate the subjects in order to make the latter believe in the legitimacy of the rule, and that they in fact do this when they need to. In some cases, however, they might not need to, viz. if the subjects spontaneously invent this ideology for the sake of their peace of mind. The consequence law cited in the previous paragraph would then *necessitate* resignation, but would not *explain* it in the cases where the subjects preempt the indoctrination by creating their own ideology. Although Cohen is acutely aware of the difficulties that epiphenomena and preemption create for the Hempelian model of explanation,[51] he is strangely unaware of the extent to which they destroy his own theory of functional explanation.

Independently of these matters of principle, there are strong pragmatic reasons for being sceptical of the use of consequence laws to back functional explanation. These are related to some basic differences between functional explanation in biology and in sociology. First, biology relies on the idea of optimal consequences, whereas sociology rests on the much vaguer notion of beneficial consequences. Secondly, biology invokes the same consequence in all cases, viz. reproductive adaptation, whereas in the social sciences the explanatory benefits differ from case to case.[52] Given this latitude in the notion of beneficial consequences, it appears likely that for any important social and historical phenomenon one may find a consequence to which it is linked in a spurious consequence law. There are so few historical instances of such phenomena as the transition from one mode of production to another, to take Cohen's major example, that little ingenuity would be required to find a spurious consequence law confirmed by them all. This objection, then, states that in the social sciences it may be difficult to distinguish between lawlike and accidental generalizations of the kind embodied in consequence laws. The objections argued in the previous paragraphs state that even a lawlike generalization may fail to explain, because of the possibility that we may be dealing with epiphenomena or preemption.

3. Intentional explanation

Intentional explanation is the feature that distinguishes the social sciences from the natural sciences. There is no point in debating whether it is also the most important method of explanation within the social sciences. Causal analysis certainly is also very important, both at the individual and at the collective level. I shall devote most of the present chapter to an analysis of intentionality, and then end with some remarks on the relation between causal and intentional analysis in the social sciences.

The conceptual network that underlies the analysis of intentionality is fairly complex. I shall start out with a table that gives the structure of the argument that follows, so that the reader at any given point in the exposition can refer back to the table and locate its place in the conceptual hierarchy.

Table 3

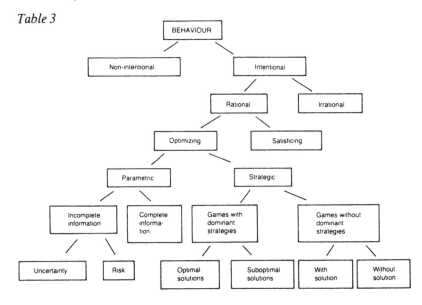

a. Intentionality

Explaining behaviour intentionally is equivalent to showing that it is intentional behaviour, i.e. behaviour conducted in order to bring about some goal. We explain an action intentionally (or *understand* it, as is sometimes said in a terminology different from the one adopted here)[1] when we are able to specify the future state it was intended to bring about. We are not, of course, explaining the action in terms of some future state: first because the explanandum cannot precede the explanans, and secondly because the intended future state may not come about at all, for a number of reasons. In particular, some intentions may be inherently unrealizable, and yet be invoked in explaining behaviour undertaken to realize them. This will prove important when discussing irrational intentions.

The general scheme for explaining intentional behaviour involves not only goals or desires, but also beliefs. An intentional agent chooses an action that he believes will be a means to his goal. This belief in turn is linked with various beliefs about factual matters, causal relations between means and ends etc. Intentional explanation essentially involves *a triadic relation between action, desire and belief*. Since the beliefs and desires are themselves in need of explanation, such intentional explanation is far from a rock-bottom analysis of behaviour. Later in this chapter I shall have something to say about the (causal or intentional) explanation of the beliefs and desires themselves.

Let us use 'reason' as a common term for beliefs and desires, and distinguish between 'acting with a reason' and 'acting for a reason'. Acting with a reason means that the actor has reasons for doing what he does, acting for a reason implies in addition that he did what he did because of those reasons. Intentional explanation involves showing that the actor did what he did for a reason. The need for this distinction is shown by cases in which the actor accidentally does something that happens to coincide with what he believes to be a way of promoting his desire. Compulsive behaviour, for example, may occasionally be adequate to the situation, but this does not make it intentional. The requirement that the actor does what he does for a reason implies that the reason is causally efficacious in bringing about the action,[2] but is not exhausted by that implication. We need to add that the reasons cause the action 'in the right way', i.e. not by a fluke.[3] We must exclude, that is, not only the 'coincidences of the first kind', in which something other than the reasons causes the action for which they are reasons, but also the 'coincidences of the second kind', in which the reasons do in fact cause the action for which they are the reasons, but do so in a non-standard way.

Intentional behaviour is essentially related to the *future*. It is action guided by a goal that is absent, not-yet-realized, merely imagined and represented. As noted by François Jacob, men can choose between unactualized possibles, whereas natural selection can choose only among the actual alternatives.[4] It has been widely noted that the capacity for deferred gratification or waiting characterizes men and distinguishes them from other animals.[5] In addition to waiting, which involves the capacity to reject favourable options in order to gain access to even more favourable ones later on, men also have the capacity for using indirect strategies, i.e. to accept unfavourable options in order to gain access to very favourable ones later. Both of these modes of behaviour crucially involve relating to the future, as does also the more complex capacity for precommitment and other strategic ways of overcoming one's own irrationality.[6]

In Ch. 2 I repeatedly invoked the idea that whenever we want to explain a pattern by its long-term positive consequences, while also imputing to it negative short-term consequences, we are implicitly presupposing the presence of a conscious decision-maker. Consciousness, indeed, may be defined as a medium of re-presentation, an inner screen on which the physically absent can have a presence and make a difference for action in the present. Operationally, consciousness can be detected through the ability to deploy indirect strategies or to wait in qualitatively novel situations. Although animals sometimes behave in these ways, they typically do so only in highly stereotyped contexts. And when some animals appear to do so spontaneously, the proper conclusion to draw is that they are indeed endowed with consciousness and the capacity to behave intentionally.[7]

It follows from this argument that the notion of unconscious intentions is no more coherent than that of a square circle. It does not follow, however, that it is impossible to make sense of the notion of the unconscious, if we conceive of it strictly as a mechanism for climbing along a pleasure-gradient.[8] It would be absurd to impute to the unconscious the capacity for waiting, for making sacrifices, for acting according to rules, etc., for these modes of behaviour all presuppose consciousness.[9] Phenomena such as wishful thinking[10] or weakness of will[11] stem from the 'pleasure principle', i.e. from the tendency to seek immediate gratification. They should be explained in terms of the 'wirings of the pleasure machine',[12] not by invoking some mysterious agency or homunculus within the person. Many psychoanalytic explanations of behaviour stem from the same misguided obsession with meaning that underlies much of functional explanation.

b. Intentionality and rationality

Can there be intentionality without rationality? Or rationality without intentionality? My main concern in this section is with the first question, but let me also say a few words about the second. It all depends, clearly, on how we define the notion of rationality. If we only mean adaptation in the sense of local maximization, we have seen in Ch. 2 that there can indeed be non-intentional rationality. But whichever way we define rationality, I believe it should be reserved for the cases in which it has explanatory power. One should never, that is, characterize a belief, an action or a pattern of behaviour as rational unless one is prepared to argue that the rationality *explains* what is said to be rational. Or, if the term is used in a non-explanatory sense, this should be made clear. The term 'rational', like the term 'functional', is often used to characterize action in a way that makes it unclear whether there is indeed an explanatory intention.

The usual way to define rational behaviour is by invoking some notion of optimization. One argues, that is, that the rational agent chooses an action which is not only *a* means to his end, but *the best* of all the means which he believes to be available. I shall argue in the next section, however, that the notions of rationality and optimality are not synonymous. The fuller characterization of rationality will have to be postponed to that section. For the present purposes it suffices to observe that rationality minimally implies *consistency* of goals and of beliefs. To drive in a wedge between intentionality and rationality, we shall have to show that there can be inconsistent desires and inconsistent beliefs.

Concerning inconsistent beliefs, I shall bring out their possibility by citing a story about Niels Bohr, who at one time had a horseshoe over his door. When asked whether he had placed it there because he believed it would bring him luck, he answered: 'No, but I am told that they bring luck even to those who do not believe in them.'[13] Rigging the story a bit, this comes out as follows:

(1) Niels Bohr believes 'The horseshoe will not bring me luck'.
(2) Niels Bohr believes 'Horsehoes will bring luck to those who do not believe they will bring them luck'.

Although the beliefs within quotation marks are consistent with each other, they cannot both be true *and be believed* (by Bohr). But a system of beliefs is consistent only if there is some possible world in which they are all true and believed.[14] If for the sake of argument we assume that Bohr was not making a joke, and that he did in fact place the horseshoe over the door because he wanted luck and believed, although inconsistently, that it would bring him luck, we have an instance of an action which is clearly irrational, yet explained intentionally.

More central, however, are the irrational desires that since Hegel have been studied intensively by philosophers and psychologists. Consistency criteria for desires can be defined analogously to those for belief: there should be some possible world in which the desire (i) is fulfilled and (ii) fulfilled through the attempt to fulfil it. The need for the first clause is obvious: it serves to characterize as irrational such desires as the well-intentioned wish that everybody earn more than the average income. The need for the second clause is shown by observing the importance of phenomena that are *essentially by-products,* i.e. the set of states that can only come about, not be brought about by deliberate action.[15] A paradigmatic case is sleep, which – after a certain point – eludes him who tries to bring it about, while mercifully coming when he has finally decided that it is going to elude him. Similarly for other 'willed absences', such as forgetfulness, indifference or non-willing (in Buddhism). There are also positively defined states, such as belief or courage, which are impossible to achieve at will, and the attempt to do so is then an irrational intention. Moreover, there are cases in which it is conceptually impossible to bring about by mere command certain states in another person, e.g. in the attempt to command the gratitude of another. Or, to take an obvious paradox, consider the command 'Be spontaneous'. This involves trying not to try, i.e. higher-order intentions, just as the Bohr paradox involved higher-order beliefs. Trying not to try, or willing the absence of will, would be consistent goals if criterion (i) alone was invoked. Since, however, they are clearly self-defeating, clause (ii) is needed.

A further example of inconsistent desires may serve as a transition to the discussion of optimality. Let us assume that a person is engaged in finding the solution to a maximization problem that in fact does not have a solution, such as 'Find the smallest real number which is strictly larger than 1'. The behaviour of this person can then be explained by the goal he has set for himself. We observe him writing down ever smaller numbers, all strictly larger than 1, and explain what he does in the light of this inconsistent plan. This not only drives in a wedge between intentionality and rationality, but also a further one between rationality and optimality. Clearly, rational behaviour in this case is not the optimizing one. What would be the (or a) rational behaviour, depends on the practical motives behind the operation. If the person is to receive a sum of money equal to 1 pound divided by the difference between the number chosen and 1, he can decide on a 'satisfactory' sum and choose the number accordingly.

c. Rationality and optimality

In this section I shall first pursue these last remarks, and then forget about them. In fact, although there are strong reasons of principle to insist on the distinctions between intentionality and rationality, and between rationality and optimality, explanation in terms of optimization remains the paradigm case of intentional explanation in the social sciences outside psychology. In economics, the theory of satisficing is recognized by many as an important theoretical alternative to optimizing, but remains insufficiently developed to serve as a basis for empirical work.

There are two reasons why the interpretation of rationality as optimality does not hold generally. These are also reasons for believing that rationality sometimes must be understood as satisficing, i.e. as finding an alternative that is 'good enough' for one's purpose rather than the 'best'.[16] First, there is *the special argument for satisficing* which derives from optimization problems without well-defined solutions, as in the case cited above. For a less trivial example, consider planning over infinite time. A planner trying to maximize consumption over infinite time may find – using the 'overtaking criterion'[17] – that there is no optimal rate of saving, since for every rate short of 100 per cent there is a higher one that is better, although the rate of 100 per cent is inferior to all others, since it implies that consumption is always postponed.[18] The rational behaviour in such cases must be to find a plan that is 'good enough', formalized in the theory of agreeable plans.[19] If in fact we observe a planner choosing a fairly high savings rate in such cases, we can explain his choice intentionally, in terms of the goal of maximizing consumption over time. Observe, however, that the explanation is not a 'single-exit' one,[20] i.e. from the assumption about the goal we cannot derive a unique observational consequence to be compared with actual behaviour. The intentional explanation must be supplemented by some causal account of why the planner set exactly that savings rate. The reason why satisficing has not emerged as a full alternative to optimizing is no doubt that such supplementary explanations have not been forthcoming, while the assumption of satisficing by itself may be compatible with a great many observed facts.

Secondly, one may invoke the *general argument for satisficing* which derives from the paradoxes of information further discussed in Ch. 6 below. Here I shall only observe that the argument requires that we do not define rationality in terms of *given* beliefs, but rather ask whether the beliefs themselves are rational. This means that to the question, 'Can this action be explained as optimizing behaviour?', we may give different answers, depending on whether beliefs are seen as constants or as behavioural variables. If we adopt a 'thin' notion of rationality, defined with respect to given beliefs, only the special argument for satisficing applies. The general argument gets its force if we adopt a broader notion of rationality, which also requires rationality in the gathering of information and the shaping of beliefs.

I now proceed to the fine structure of optimizing behaviour. The first and simplest case is what I shall call parametric rationality, i.e. rational behaviour within an environment that the agent (perhaps wrongly) assumes to be made up (a) of natural objects ruled by causal laws and (b) of other agents who either are such that their behaviour makes no difference to him or, if it makes a difference, are assumed to be less sophisticated than he is himself. The last clause implies that the agent thinks of himself as a variable and of all others as constants; or if he thinks that the others are adapting to their environment, he believes himself to be the only one to adapt to the others' adaptation, and so on. This takes care of the 'parametric' part of the notion of parametric rationality. The rationality part implies that the agent tries to do as well for himself as he can, given his beliefs about the world. In many cases this goal can be represented by some objective function, which may be real or notional. A real objective function is one that the agent consciously sets out to maximize, as when one tries to explain the behaviour of a firm by assuming that it (i.e. the manager) tries to maximize profits. A notional objective function is a utility function that may be constructed as a convenient shorthand expression for an important subclass of consistent preference orderings.[21] The standard case of parametric rationality is maximization of an objective function within given constraints, but it should be clear from the remarks above that the full definition is somewhat more complex.

Even if the actor assumes the environment to be a parametric one, he may have less than complete information about it. In this case we must make a basic, but contested, distinction between risk and uncertainty (see Appendix 1 for a fuller discussion). There is risk when the agent has quantifiable degrees of belief, or 'subjective probabilities', about the various possible states of the world. Rationality in this case implies that the agent should maximize the expected utility associated with the various courses of action, i.e. an average of the utilities that will be realized for different states of the world. Uncertainty, on the other hand, arises when the agent is not able to specify any numerical probabilities, not even within a range of lower and upper limits. Or, even more fundamentally, he may not even be able to specify a complete set of the possible states of the world, let alone their probability. In the first of these cases, there is an important theorem stating that the agent can rationally take account only of the best possible and the worst possible outcomes associated with each course of action, but that he cannot rationally decide how much weight to give to the one or to the other. This will depend, for example, on such character traits as 'optimism' or 'pessimism', so that the final decision can only be explained causally. In the second case, of course, there is even less scope for rational decision-making. In the first variety of uncertainty, we may at least exclude some courses of action as unambiguously inferior, viz. those whose best-consequence is worse than the worst-consequence of some other alternative. In the second variety, however, even this kind of comparison cannot be carried out, since we do not know the full range of possible states of the world and thus of possible outcomes of the various courses of action.

Strategic rationality is defined by an axiom of symmetry: the agent acts in an environment of other actors, none of whom can be assumed to be less rational or sophisticated than he is himself. Each actor, then, needs to anticipate the decisions of others before he can make his own, and knows that they do the same with respect to each other and to him. The strategic approach to human behaviour is formalized by game theory, which might more appropriately have been called the theory of interdependent decisions.

To bring out the proper place and contribution of game theory in the analysis of social interaction, we may observe that social life is constituted by four, interlocking, sets of interdependencies. First, *the reward of each depends on the choice of all*, through general social causality. Secondly, *the reward of each depends upon the reward of all*, through envy or altruism. Thirdly, the *decision of each depends upon the decision of all:* this is the specific contribution of game theory. Lastly, *the preference structure of each depends on the actions of all*, through socialization and similar mechanisms. All these statements are, of course, intended to cover the most general case; in particular cases they may fail to hold good.

The full structure of game theory is fairly complex, and no attempt is made here to sketch even the outlines.[22] I shall say a few words, however, about the basic distinction between cooperative and non-cooperative games, and then point out some important cases within the latter category. Cooperative game theory assumes that groups of agents can act together against other groups, and does not inquire into the likelihood, or the conditions for, such cooperation to come about. This implies, in my view, that the theory cannot have great explanatory power, although it may be very useful for the purpose of normative analysis.[23] Non-cooperative game theory is more satisfying in this respect, since it postulates nothing but individually rational decisions. Cooperative game theory requires foundations in non-cooperative theory, showing that the decision to cooperate can be a move within a non-cooperative game. Or, alternatively, one may attempt to show that there are (non-strategic) mechanisms that tend to bring about the 'solution' to the cooperative game, through some kind of 'invisible hand'.[24] To assume that the cooperative behaviour will be brought about simply because it is collectively optimal, is to fall victim to functionalist thinking.[25]

Within the class of non-cooperative games, one variety is theoretically trivial, although important in applications. These are the games in which each participant or player has one course of action, or strategy, that is his best choice regardless of how others choose. The Prisoners' Dilemma set out in Ch. 2 above has egoism as a *dominant strategy* in this sense. In this we saw that it was rational for each individual to act in a way that, when adopted by all, is disastrous for all. Although in this game the reward of each is affected by the decision of all, the decision of each can be taken independently of the decisions of all. Consider, by contrast, a situation usually called the 'Assurance Game'.[26] In the language of Ch. 2, this game is defined by postulating that all players rank the alternatives as follows:

Assurance Game Preferences: 1. (A,A) .2. (E,A) .3. (E,E) .4. (A,E).

Here egoism is no longer dominant, for if all others behave altruistically, this is also what the individual agent prefers to do. The optimum (A,A) is individually stable. It is not, however, individually accessible, for if others behave egoistically, then the individual will do so too. There is no desire to be a free rider, although there is still the fear of being a 'sucker'. Even though the game does not have dominant strategies, it has a solution, viz. (A,A). I shall not spell out the precise definition of the solution concept in non-cooperative game theory, but informally speaking it consists of the set of strategies towards which rational and perfectly informed players will tacitly converge. If the solution is made up of dominant strategies, only lack of individual rationality can block it. In games without dominant strategies, lack of information may block the solution. Thus in the Assurance Game the optimum will not be brought about if each player is unaware of the preferences of the others, believing perhaps that they are as in a Prisoners' Dilemma; or if each believes that the others have Assurance Game preferences, but also believes that the others do not know this about each other, and so on. For the solution to be realized, shared Assurance Game preferences is not enough; the 'sharing itself must be shared'.[27]

This implies that because of lack of information one may get an outcome that is *worse for all* than another feasible outcome. Such an outcome is called 'Pareto-suboptimal'. It is clearly a perverse and disturbing feature of social interaction. It is also present in the Prisoners'

Dilemma, though for a different and deeper reason. Suboptimality in the Assurance game stems from information failure, in the Prisoners' Dilemma from coordination failure. One may ask whether the information failure of the Assurance Game will not be the rule, since the solution depends on stringent information requirements that are not likely to be fulfilled in actual cases.[28] In the extreme case we have been setting up, this may well be true. But we have been using the extreme model only to clarify the exposition, and it is quite possible to modify it so as to make it more realistic. In particular, one may construct a more powerful framework that permits choice to be contingent upon various *proportions* of other participants choosing one way or another, rather than making it depend on *all others*.[29]

When one is trying to explain observed sub-optimalities in social life, it may be hard to ascertain whether they are due to information failures or to coordination failures.[30] One may have reason to believe, for instance, that most people will prefer a state in which no one engages in pollution, tax evasion and the like to one in which everybody does so, and yet one observes that pollution and tax evasion are rampant. The difficult explanatory task is to determine whether the preferences are as in a Prisoners' Dilemma or as in an Assurance Game. The problem also has important practical implications, since the techniques for overcoming suboptimality differ radically in the two cases. In particular, one may argue that an important task of leadership is to provide the information that enables people to converge on a behaviour for which they all have a conditional preference. With unconditionally altruistic preferences there is no need for leadership, and with unconditionally egoistic preferences one needs coercion rather than leadership to bring about the collective optimum.[31]

Even more perverse than suboptimalities due to information or coordination failure are games without a solution, i.e. situations such that there is *no* behaviour that is strategically rational. These games fall in two classes. First, there are the games in which there is no set of strategies that are individually stable, i.e. no set of strategies such that no one can do better for himself by breaking out. The technical term for a set of strategies with this property is an *equilibrium point*. A simple game without an equilibrium point is the following. Each actor has to think of a number and write it down secretly. When the numbers are compared, the player who has written down the greatest number gets from each of the

80 MODES OF SCIENTIFIC EXPLANATION

others a sum equal to the difference between his number and the one written down by the other. Clearly, any player can always do better by writing down a greater number. Hyperinflation may be a more substantive example of an interaction structure with similar properties.

Secondly, there are games with more than one equilibrium point, none of which will emerge as the focal point for tacit convergence. The Assurance Game has two equilibrium points, (A,A) and (E,E), of which the former emerges as the solution because everybody prefers it to the latter. But there are also games with multiple equilibria of which none can be singled out as the solution. Bargaining and bilateral monopoly are typical cases. Consider also the following variation on the themes presented above:

Chicken Preferences: 1. (E,A) .2. (A,A) .3. (A,E) .4. (E,E).[32]

Here the situation is such that each has an incentive to use E if all others use A, and vice versa. It is easy to see intuitively that this makes for instability, and for everybody chasing everybody else in an endless pursuit. More formally, the game has the following equilibrium points. With more information about the payoff structure of the game, we can determine a proportion p of the players such that if p choose A and $1-p$ choose E, we have an equilibrium point in which no one can improve their outcome by switching to another strategy. In fact, this gives us a large number of equilibrium points, as there are many permutations of the players compatible with these proportions. In addition, there is the equilibrium point in which each actor chooses A with probability p and E with probability $1-p$. This is a *mixed strategy,* already referred to in Ch. 1 above. On rational grounds, there is no way in which the actors could choose between these equilibria. On psychological grounds, it can be argued that the mixed strategy stands out as a *focal point,*[33] being qualitatively different from all the others. But there may always be doubt about the psychology of the other players, and in any case there are other no-solution games that do not have any psychologically salient equilibrium points.

When playing a game with no rational solution, each player has to make some assumption about what the others are going to do, and then act to maximize his reward on that assumption. The assumption cannot be a rational one, in the sense of being derivable from the hypothesis that the others are as rational and well-informed as himself. It has to draw on some psychological knowledge or belief, either about the human players specifically or about human beings generally. The situation becomes intolerable, however, since each player, while trying to outguess the others, knows that they are trying to outguess him. Each player is rational and knows the others to be so, and to know as much about the situation as he does himself, and yet he has to treat them as causally determined, knowing that they treat him similarly. This is the reintroduction of parametric thinking within strategic rationality. Intentional explanation is not sufficient in such cases. It may exclude some courses of behaviour as clearly irrational, but only causal theory can then narrow down the remaining possibilities to a uniquely determined outcome. To be sure, the situation is not 'objectively indeterminate' in the sense of quantum mechanics,[34] but it can be made determinate only by supplementing the intentional model with a causal theory.

An example may be helpful here. There is a large literature on the rationality of political participation: would a selfish and rational person bother to vote? Surely he would know that his chance of having an influence on the outcome is virtually nil, and smaller in any case than the probability that he will be hit by a car and killed on his way to the voting place. In nationwide elections this is certainly true, and then we can only explain that people vote by assuming that they are not narrowly self-interested, or not moved exclusively by the consequences of what they do. But in smaller electorates, a person might reason as follows.[35] 'Surely other people will see that it is irrational for them to vote, except for a small number of "ethical voters". These, however, will be so few that my voting actually could make a difference, so it is rational for me to vote. But wait! Others might argue in the same way to themselves, and turn out to vote in so large numbers that it may not be rational for me to vote after all. So I'd better stay home. But wait again! The others might think in the same way, making it rational for me to vote after all.' And so on. The infinite regress may at some point be broken by action, i.e. by the person going to vote or making a conscious decision not to vote. One might then argue that the person has formed an implicit or explicit estimate of how others are going to act and done as well as he can for himself on that assumption. But what if the person goes on deliberating until the polling station is closed? His predicament is to some extent like that of Buridan's ass, though to me it represents a far deeper anomaly in the theory of rational choice.

We may use these game-theoretic notions to gain further insight into the difference between intentional and functional adaptation. First, consider the Asssurance Game. Animals sometimes seem to behave in a way corresponding to this situation, e.g. by adopting the strategy of 'swamping the appetite of predators'.[36] It is in the nature of the case that this behaviour may be effective when adopted by many prey simultaneously, but the first one to try it out would hardly have any reproductive advantage. The behaviour, then, is not individually accessible, but there are cases in which it is individually stable, because 'there is safety in numbers'. But since individual accessibility is what natural selection is all about, one cannot explain the presence of such behaviour by arguing that it forms an 'evolutionary stable strategy'. In human interaction, by contrast, individual stability can explain the presence of a behavioural pattern if the relevant information requirements are fulfilled. If all know that a certain arrangement is best for all, and if there is no temptation to defect, and all know this, then that arrangement will emerge spontaneously even if there is no individual incentive to contribute unilaterally.

Secondly, consider cases of Chicken in the animal realm, such as 'the logic of animal conflict'.[37] Here we assume that there are two behavioural variants among the organisms in a population: the 'hawks' and the 'doves', so-called because of their behaviour when they meet and fight. We assume that when dove meets dove, each of them has a 50 per cent chance of winning 50 'evolutionary units' (a measure of reproductive success) and of losing 10; that when dove meets hawk, the former will neither win nor lose and the latter win 50 with certainty; and that when hawk meets hawk they have equal chances of winning 50 and losing 100 units. It is easy to see that in a population of doves, the lone hawk will have a reproductive advantage, and vice versa. One can also show that the equilibrium is a stable polymorphism in which there are 5/12 doves and 7/12 hawks. In human interaction one could also, as argued above, have the possibility of all individuals using a mixed strategy, acting on each occasion as hawks with probability 7/12 and as doves with probability 5/12. This, however, is implausible in the animal realm.[38] A 'switch gene' would have an initial disadvantage in any population except one that had attained the 5:7 ratio, and even here it would not confer any advantage: it would be as good as either pure strategy, but no better. The moral is the same as in the preceding example: animal adaptation is strictly individualistic, and could never favour behaviour that is beneficial only when adopted by all. Intentional beings, by contrast, can choose in terms of their expectations, including their expectations about the expectations of others. In the social sciences, then, there is more scope than in biology for explanation of behaviour in terms of collective benefits, assuming, however, individual stability. That something is good for all may explain why all do it, assuming that it is not even better for each to be the only one not to do it.

Let me conclude this survey of rationality theory with a few remarks about the anomalies we have encountered.[39] First, the very notion of irrational intentions goes against the grain of many recent discussions, which tend to see the ideas of intentionality and rationality as synonymous. Secondly the special and general arguments for satisficing break the link between rationality and optimality. Thirdly, the varieties of decision-making under uncertainty induce scepticism about the power of rationality theory as a guide to action. And lastly, the existence of games without any solution shows that individual rationality may break down when the interaction structure is sufficiently perverse. To be sure, these last conclusions will be contested. Some will say that one may always form some anticipations or subjective probabilities about the behaviour of others or the state of the world, and maximize expected utility accordingly. In Appendix 1 the reader will find an attempt to refute this objection. Others will argue that even under uncertainty or in a game without a solution one may act with some rationality, viz. by satisficing. To be precise, the notion of 'maximin behaviour' – choosing the course of action with the best possible worst-consequence – can be seen as a variety of satisficing. But observe that in games without an equilibrium point, like the hyperinflation game, and in decisions under radical uncertainty, which extends even to the range of possible states of the world, the notion of maximin behaviour is not well-defined. I conclude that the anomalies are genuine difficulties, not to be resolved in Procrustean fashion.

d. Intentionality and causality

Game theory studies what one may call *intentional interaction between intentional beings*. There also occurs, of course, purely *causal interaction* between intentional agents. This takes place when each agent acts upon unjustified assumptions about the behaviour of others, e.g. when each agent believes that he is the only one who is adjusting to the environment, whereas all others merely follow habit or tradition. Or again, to repeat a phrase used above, when each agent looks at himself as a variable and all others as parameters, seeing himself as uniquely *pour-soi* and all others as *en-soi*.[40] This mode of interaction can bring about dramatic unintended consequences. Indeed, *counterfinality*, to use Sartre's term,[41] stands with suboptimality and games without solutions as one of the main contradictions of social life.[42] For an example of how intentions can be thwarted in this manner, consider the cobweb cycle from economic theory.[43] Here each producer has to decide in year t on how much to produce in year $t+1$. If acting naively, he will decide on the volume that maximizes his income in year $t+1$ on the assumption that prices will remain the same as in year t. With low prices in year t this will bring about a small volume in year $t+1$, leading to high prices in year $t+1$, which bring about a large volume in year $t+2$ and so on. The assumption of constant prices implies an assumption that all other producers will market the same volume as in the current year, and that the adjustment of the individual makes no difference to the price level. The second part of the assumption is justified in a perfectly competitive market, the first is not.

This provides a paradigm for many cases of analysis in the social sciences: *intentional explanation of individual actions together with causal explanation of the interaction between the individuals*. First we must 'understand' why – i.e. for the sake of what goal – the actors behave as they do; and then we must 'explain' why, behaving as they do, they bring about what they do. Whether the phenomenon to be explained is a business cycle, a presidential campaign, geographical mobility or technical change, one should try to decompose the explanation into these two stages. Simply to postulate causal relationship between macro-variables will not do. We may observe as an empirical regularity that upswing will follow downswing in the business cycle, or that certain patterns of income distribution bring about certain patterns of migration, but we have *explained* nothing until we can show (i) how the macrostates at time t influence the behaviour of individuals motivated by certain goals, and (ii) how these individual actions add up to new macrostates at time $t+1$.

There is, however, also 'sub-intentional causality', i.e. causal processes shaping the beliefs and desires in terms of which actions can be explained intentionally. Let me first point out that when we seek to explain desires and beliefs, it is not obvious that only causal explanations are available. One may in fact choose one's desires, through 'character planning',[44] which then provides an intentional explanation of why the desires are what they are. But then, of course, the question arises again at one further remove, and presumably there will be some stage in the regress at which only a causal explanation is forthcoming.[45] Similarly, but more controversially, it may be possible to choose one's beliefs, in the sense of deliberately adopting certain beliefs because it is useful to hold them, quite independently of whether the agent believes them to be true.[46] It is probable, though, that the deliberate choice of character traits and belief systems is a rare phenomenon. In any case I shall largely limit myself to the causal explanation of desires and beliefs.

Desires are shaped, predominantly, by socialization. This does not mean that one is socialized into desiring some particular good, regardless of costs. Rather the idea is that one learns from socialization how to trade off different goods against each other. A criminal, say, is not one who has developed within a criminal sub-culture that deprives him of any choice of career at all. Rather he is a person who, when making this choice, tends to attach greater importance to some consequences than to others. He may, for instance, be a risk-preferrer rather than risk-averse; he may give greater weight to short-term than to long-term consequences, and so on. To say this is not to say that he will take *any* risk, or that he prefers *any* short-term benefit to *any* long-term gain. Nor is it to say that it will be impossible to affect his choice by modifying the reward structure, e.g. by more severe penalties or increased probability of detection. It may be to say, however, that this way of changing his behaviour may be relatively expensive, and that working directly on his environment may be a more efficient method. To sum it up: one should not seek in socialization the direct spring of action, only the cause of certain preference schedules that, in a given environment, may lead an action to be preferred to the feasible alternatives. Adding a new tier to the two-tiered structure indicated above, we may formulate the following slogan: first a causal explanation of desires, then an intentional explanation of action in terms of the desires, and finally a causal explanation of macro-states in terms of the several individual actions.

Socialization, however, is not the only causal mechanism that shapes our desires. There is also 'adaptive preference formation', summed up in the story of the fox and the sour grapes.[47] This comes about through what I referred to above as the wirings of the pleasure machine, e.g. by the tendency to reduction of cognitive dissonance. Other causal processes shaping the desires include the production of dissonance, e.g. through the perverse drive for novelty,[48] or such imitative phenomena as conformism and anti-conformism. These processes suggest the idea of a *general sociological theory*, in which preferences and desires are explained endogenously as a product of the social states to the generation of which they also make a contribution, as explained towards the end of the preceding paragraph. This theory – which, needless to say, in the present state of the arts appears to be light-years away – would include (i) the explanation of individual action in terms of individual desires and beliefs, (ii) the explanation of macro-states in terms of individual actions, and (iii) the explanation of desires and beliefs in terms of macro-states.

But there is more to be said about the causal processes of preference formation. The mechanisms discussed in the preceding paragraph are predominantly 'hot', i.e. they explain the formation and change of preferences through the drives and pleasure-wirings of the individual. But desires may also be shaped by 'cold' mechanisms, i.e. by cognitive distortions somewhat akin to optical illusions. Thus the relative attractiveness of options may change when the choice situation is reframed in a way that, rationally, should have no impact on the preferences. Tversky and Kahneman cite an example from L. J. Savage of 'a customer who is willing to add $ X to the total cost of a new car to acquire a fancy car radio, but realizes that he would not be willing to add $ X for the radio after purchasing the car at its regular price'. And they go on to add that 'Many readers will recognize the temporary devaluation of money that makes new acquisitions unusually attractive in the context of buying a house'.[49]

Going from desires to beliefs, we also find hot as well as cold processes. 'Hot' belief formation includes such phenomena as wishful thinking, rationalization and self-deception, i.e. cases in which our beliefs about what the world is like is shaped by our desires concerning what it should be like. Similarly, 'cold' errors include those which result from cognitive failures, as when 'an individual judged *very* likely to be a Republican but rather *unlikely* to be a lawyer would be judged *moderately* likely to be a Republican lawyer',[50] as if the probabilities were additive rather than multiplicative.

But of course not all belief formation is erroneous or biased belief formation. There is also belief formation that in some sense is correct or rational, not necessarily resulting in true beliefs, but at least in justified beliefs. (I here use the term 'justified' in the strong sense that the person must have the belief for good reasons, and not merely that there should be good reasons for the belief he has. The latter, in fact, is compatible with wishful thinking, if his desires lead him to hold beliefs for which he also has good reasons, even though those reasons in fact played no role in the belief formation. The strong sense of 'justified', that is, is needed to exclude 'coincidences of the first kind'.) People with the ability to form rational or well-grounded beliefs are often said to have *judgement*, a crucial quality in many walks of life. Saying that a person holds a certain belief because he has good reasons for it is somewhat like offering an intentional explanation of the belief, yet not quite. We are not saying that he has chosen the belief, but that in having the belief he shows himself to be rational. This holds to an even higher degree if the person goes beyond having beliefs that are rational relative to the available evidence, and goes on to acquire more evidence if he feels it is needed.

I am baffled by this anomalous form of explanation, which does not fall neatly into the intentional or the causal variety. A similar problem arises in the formation of desires. Many of the mechanisms that shape our wants lead to heteronomy, as when someone always wants to do the opposite of what the majority is doing – not because he has a meta-preference for such wants, but because he is a contrary sort of person. Such persons do not have their moving principle within themselves, but behave more like appendages to the social structure. And there are many other mechanisms of this kind. By contrast, of an autonomous person we can say that he possesses the desire, rather than that the desire possesses him. Unfortunately I do not know how to define the notion of autonomy, although I feel certain that there are autonomous persons. To define autonomy through the idea of intentionally chosen wants gives both too much and too little: too much because there may be non-autonomous meta-desires, too little because autonomy may exist without this self-reflective attitude. But assuming that we knew how to define or at least how to recognize autonomous persons, there would still be a conceptual problem as to what kind of explanation we are proposing when we say that a person acts on an autonomous desire.

In fact, the notion of rational behaviour relative to *given* (and consistent) desires and beliefs is an extremely thin one. In addition to this formal rationality we want people to have substantive rationality, in the twin forms of *judgement* and *autonomy*. The formula of rational-cum-intentional explanation of action in terms of desires and beliefs, supplemented with causal explanation of the beliefs and desires themselves, may turn out to be misleading or superficial. If people are *agents* in a substantive sense, and not just the passive supports of their preference structures and belief systems, then we need to understand how judgement and autonomy are possible. This, in my view, is the outstanding unresolved problem both in philosophy and the social sciences.

Part II
Theories of technical change

Introduction to part II

In the remainder of this book I survey some theories of technical change from the vantage point provided by the foregoing account of scientific explanation. I begin with the neoclassical orthodoxy, which tends to explain technical change as just another instance of maximization under constraints. I believe I say nothing new when arguing that neoclassical theory is at its best when dealing with static settings, including intertemporal equilibria. The extension of the theory to the dynamic problem of innovation is, however, problematic. Basically, the difficulty is the following. Although it is certain that innovation at any given time is constrained by what is scientifically and otherwise possible, these constraints cannot enter into the explanation of innovation unless they are known to the innovator. Although the innovating entrepreneur may be assumed to maximize profits, we cannot impute to him the knowledge of the feasible set that would enable us to explain his behaviour on the standard model of parametric rationality.

Although the neoclassical theory dominates contemporary economics quite generally, extending its influence even to Marxist economic analysis,[1] it is probably fair to say that within the study of technical change it is rivalled in importance by the Schumpeterian tradition. Schumpeter stressed the irrational side of entrepreneurial innovation: the dynastic ambitions and the excessive optimism that, even if disastrous for the majority of entrepreneurs, is socially useful. He also departed from the orthodox view when suggesting that the monopolistic elements of capitalism that were generally condemned for their allocative inefficiency were indispensable for growth and innovation. Although Schumpeter's theory proved resistant to formal modelling, it has exercised a largely subterranean influence in the shaping of the way in which economists think about innovation. He was very far from being an economist's economist, yet understood better than most economists the nature of their subject matter: capricious human beings.

A third strand of thought is taken up in the chapter on evolutionary models. These are largely evolutionary in a loose analogical sense, unlike the accounts that have been proposed by ethologists of the emergence of tool behaviour in animals. I begin the chapter with a brief exposition of these literal evolutionary models, and go on to three versions of models that in some sense are based on the evolutionary metaphor. First, I draw attention to the work of Eilert Sundt, a Norwegian sociologist who in 1862 proposed what is probably the first Darwinian model of technical change. Secondly, I discuss at some length the important work of Richard Nelson and Sidney Winter, who for more than a decade have been engaged in working out a full-scale alternative to the neoclassical orthodoxy, drawing in equal amounts on Schumpeter and Herbert Simon. And lastly, I discuss the rival heresy of Paul David, who argues that technical change is the aggregate result of 'learning by doing' at the micro-level.

The last chapter, on Marxist theories of technical change, will be felt, perhaps, to be disproportionately long. Partly this is because my own personal interest and competence are higher in this area than in the others. Partly the emphasis given to Marxism also reflects the emphasis given to technical change in Marxism. Marxism is an immensely influential and intellectually challenging theory of world history, in which technical change – the development of the productive forces – plays a major explanatory role. Moreover, Marx studied in great detail the mechanisms of technical change under capitalist relations of production, suggesting several important approaches that have only recently been further opened up.

Let me set out informally what these theories are trying to explain. More formal and precise statements are given in the following chapters, but there is a common background to the theories that can be stated here. Basically, they are concerned with explaining *the rate and the direction of the change in technical knowledge*. Assuming that technical knowledge can be measured cardinally, the term 'rate of change' should not pose any difficulty. The term 'direction' refers to the factor-bias of the change, e.g. whether it saves on labour, capital, energy, etc. These are not the only aspects of technical change one might want to explain. One might also, for instance, try to account for the *location* of technical change, i.e.

whether it occurs primarily in industry, agriculture, mining etc. Or again, one might focus on the relative importance of *product-innovation* and *process-innovation*, i.e. of innovations that lead to new consumer products and innovations that permit cheaper production of existing products. I shall focus, however, on the two variables singled out above, in order to facilitate comparison between the theories and to eschew the more substantive problems which I have no competence to discuss.

The production of new technical knowledge I shall call *innovation*. I distinguish it, first, from *invention*, which is the generation of some scientific idea, theory or concept that may lead to an innovation when applied to a process of production; secondly, from *diffusion*, which is the transfer of an existing innovation to a new context; and thirdly from *substitution*, which involves change in the production process on the basis of existing technical knowledge. In most cases I shall be concerned only with innovation and substitution. Although science certainly is an important determinant on the 'supply-side' of innovations, I shall largely take it as being given exogenously. True, the rate of technical change is influenced both by the rate of scientific change and the rate of transformation of inventions into innovations, and it would appear that both of the latter are shaped to some extent by economic processes. But I shall only consider the rate of transformation, since in the short and medium term the rate of invention may be taken as given. As for diffusion, it falls largely outside the literature I am discussing. I should point out, however, that there often is an element of innovation in diffusion as well, since a method often has to be adapted to the new context.[2] Similarly, the distinction between substitution and innovation is far from watertight, for if a firm is operating with one factor combination it can rarely switch effortlessly to another.

If innovation is a change in technical knowledge, we have to know what that knowledge is. The following sketch of the structure of technical knowledge at a given point of time is not completely general. It is biased towards the neoclassical approach (and the 'production function'), and less compatible with the evolutionary models of Nelson and Winter. Yet it is sufficiently general to provide a background for most of the analysis in the following chapters.

The structure of technical knowledge is best understood as a three-tiered one. First, there is what I shall call a *practice*, which is a particular combination of factors of production used in a specific process. Secondly, there is what I shall refer to as a *technique*, i.e. a set of practices that permit some degree of substitution between the factors, so that one can switch from a practice using much of one factor and little of another, to one that uses more of the latter and less of the former. Thirdly, there is the available *technology*, by which I have in mind all known techniques. To illustrate, consider the production of nitrogen fertilizer.[3] The traditional technique involves the production of hydrogen through electrolysis of water, but there is also available a more recent technique involving the extraction of hydrogen from oil. Together these techniques make up the technology for producing nitrogen fertilizer. Within the traditional technique, one has the choice between several practices, e.g. between using more capital (electrolytic cells) and less energy (electric power) and using less capital and more energy. Similar substitution possibilities may be available within the new technique.

This way of looking at the production possibilities is somewhat non-standard. The more usual perspective is to distinguish only between two levels, e.g. between technique (corresponding to what I call practice) and technology. It is true, in fact, that often the distinction between what I call technique and technology is not so easy to make as in the example above. Further, the choice between (what I call) techniques may also involve substitution, but at a higher level of aggregation. Thus in the same example the switch from the traditional to the more recent technique might involve, conjecturally, substitution of labour for capital, where 'capital' means some aggregate of the many and qualitatively different capital goods. There are good reasons, however, for refusing the last move. The 'capital controversy', to which reference was made in Ch. 1 above, has shown, conclusively in my view, that in many cases aggregate capital is not a well-defined notion. Assume, in my terminology, that each technique consists of a single practice. It has then been demonstrated that for the several techniques to be subsumable under a single technology ('production function') which permits substitution of labour for aggregate capital, one has to postulate fairly specific conditions that cannot be assumed to be universally fulfilled. The nature of these conditions is spelled out in Ch. 7 for one simple case. Presumably the aggregation problem would be even more difficult if the distinction between practice and technique is also made. I believe, then, that in the general case the full three-tiered structure is needed, although there may be particular cases in which a simpler two-tiered model will do.

It is characteristic of neoclassical economics that it collapses the three levels into two. It is usually thought that Marx, simplifying even more, assumed that the technology at any given point of time consists of a single practice, so that there is no room whatsoever for choice between practices or between techniques. This is what is usually refered to as the assumption of 'fixed coefficients of production', or 'the Leontief technology'. In Ch. 7 I discuss whether Marx really held this view, and conclude that while he (wrongly) denied intra-technique choice, he did recognize inter-technique choice.

It follows from this conceptual discussion that one ought perhaps to distinguish between technical change and technological change, the former being an improvement of an existing technique and the latter an addition of a new technique to the existing spectrum. I shall not do this, however, but speak throughout of 'technical change'. This will reflect the fact that I shall often be concerned with the more narrow sense of the term, i.e. intra-technique improvement. In other cases the discussion will be so general that it applies to both the narrow and the broad senses. This again reflects the fact that in most of the literature the distinction between the two senses is not made.

4. Neoclassical theories

A. The production function

The basic neoclassical tool for the study of technology and of technical change is the notion of a *production function*.[1] Here and later I shall conceive of a production process as one in which a single homogeneous product is the outcome, disregarding the case of 'joint production'.[2] On this assumption we may look at production as a process with many inputs, or 'factors of production' and one output or product. Conceptually, both inputs and outputs may be understood as *points* or *flows,* and the production process may correspondingly be understood in various ways. I shall limit myself, however, to the simple 'point input-point output' case, in which we assume that at one point of time the inputs enter the factory while the output becomes available at some later point in time. The production function specifies a quantitative relation between inputs and outputs: given the quantities of the various inputs, a certain amount of final product will be generated. There is, of course, also the possibility that less will be produced, because of inefficiency, but I shall assume that there is one output that is invariably produced, given the inputs.

The causal structure of the production process involves transformation of (i) a raw material using (ii) energy, (iii) labour, and (iv) produced means of production. All production requires raw materials and energy. In addition it requires either labour or means of production, ultimately produced by labour. In the production process some inputs are completely used up; these normally include raw materials, energy and labour. The produced means of production may or may not be completely used up. We may assume for convenience that the duration of a process of production corresponds to the productive life of the produced means of production, so that all inputs are completely used up in the process. This involves some substantive simplifications, but I shall try to offer only robust conclusions.

Given this background, the general form of the production process is

$$q = f(x_1, x_2, \ldots x_n)$$

Clearly, this is too general to be helpful. To be able to say something more specific, one must impose some structure on the variables and on the function. I shall use the standard neoclassical approach[3] and assume that there are just two inputs: labour and aggregate capital. One rationale for using this objectionable representation is that we may think of the discussion as limited to the cases in which it is unobjectionable. More interestingly, many of the conclusions may be transferred, *mutatis mutandis*, to the case of *n* qualitatively different capital goods. And finally, as I am mainly expounding neoclassical views, I can hardly avoid using their apparatus even when it suits the facts badly. So we now posit

$$q = f(K,L), \tag{1}$$

where 'K' stands for aggregate capital and 'L' for aggregate labour. To impose a structure on the function f, we assume (neoclassically) that it is continuous and continuously differentiable in both variables. This implies, for example, that the notion of a 'marginal product' (of capital or of labour) always makes sense. In Ch. 7 below we shall see that the notion of fixed coefficients of production can also be represented by a production function, which, however, does not have well-defined marginal products. We also assume that there are decreasing marginal products, meaning that if one factor is held constant and the other steadily increased, each extra unit of the latter gives a steadily decreasing increase in the total product. The marginal product may or may not ultimately become zero or even negative. Moreover, one assumes that there are constant returns to scale:

$$f(aK, aL) = a \cdot f(K,L), \text{ for all non-negative } a \tag{2}$$

With these assumptions, one has the following theorem:

$$f(K,L) = f_K(K,L) \cdot K + f_L(K,L) \cdot L \tag{3}$$

In words, the theorem says that if we take the marginal product of capital multiplied by the total amount of capital plus the marginal product of labour multiplied by the total amount of labour, this exactly exhausts the total product. (For convenience, non-mathematical readers may think of the marginal product of a factor as the extra amount of product created by the addition of one extra unit of the factor, given that the pre-existing amounts were K and L respectively. This simplification only makes sense, however, if the units are small compared to the total amounts employed.) As was briefly observed in Ch. 1 above, the assumption of constant returns of scale and equation (3) can serve some dubious theories of income distribution, but this is beside the point here. In the present context it is more important to observe that the assumption of constant returns permits a convenient representation of the production function by means of *unit isoquants*. An isoquant is defined as the locus of factor combinations that give the same output, so that (K_1, L_1) and (K_2, L_2) are on the same isoquant if and only if $f(K_1, L_1) = f(K_2, L_2)$. With constant returns to scale, all the information in the production function can be conveyed by the isoquant corresponding to one unit of output, since all other processes are simple multiples of this. The unit isoquant in Fig. 1 is drawn convex to the origin, corresponding to the idea of decreasing marginal product. The economic theory of production – as

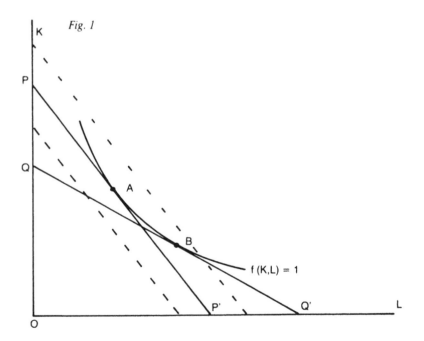

Fig. 1

distinct from the technical aspect – is concerned to determine which of the factor combinations will be realized. Neoclassical theory assumes that this is decided by rational choice, and that the entrepreneur (in a capitalist economy) chooses the factor combination that maximizes his net revenue or profit. If we assume that the prices of the factors of production and of the product are given (the assumption of perfect competition), then we can explain his choice in a particularly simple manner. In Fig. 1 each of the straight lines is an *iso-cost curve,* i.e. the locus of input combinations with the same total cost, given the factor prices. The slope of these lines corresponds to the factor price ratio. The entrepreneur will then choose the point on the unit isoquant that lies on the lowest iso-cost curve, i.e. the point A of tangency between the isoquant and the fully drawn line PP'. Similarly, a change in the factor price ratio, leading to a set of iso-cost curves parallel to QQ', will make the entrepreneur switch from A to B. This is *substitution of labour for capital* following a rise in the price of capital relative to that of labour.

Basically, this is all there is to the static neoclassical theory of production. To this simple theme an infinity of frills and variations may be added, but the basic logic of rational choice and price-induced substitution is the same. We may note that the simple model sketched above rests on three assumptions. The most important, perhaps, is the behavioural postulate of profit maximization. In later chapters we shall see how various evolutionary and Marxist theories deny this postulate – either by assuming that the choice between the possible practices follows a different logic or by denying that there is any choice to be made. Almost equally important is the assumption embedded in the notion of the production function, a compact representation of the existing technical possibilities and constraints. All points on the unit isoquant are assumed to be equally accessible to the firm, and the point corresponding to the actual practice is in no way privileged. Also, one can move along the unit isoquant in response to price changes without the isoquant itself being affected by the moves. In Ch. 6 we shall see that these assumptions have been challenged. Finally, the model rests on the institutional postulate of perfect competition. This is the least central of the assumptions, and neoclassical theory is perfectly able to handle the complications that arise when it is dropped. In fact, most of the 'frills and variations' referred to above are explorations of choice of techniques under imperfect competition.

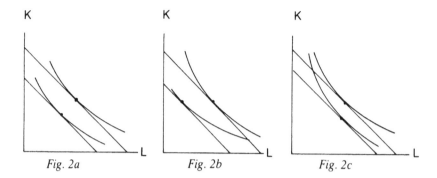

Fig. 2a Fig. 2b Fig. 2c

Technical progress, quite generally, is defined as a shift of the unit isoquant towards the origin. Following W.E.G. Salter,[4] we can define the rate or the extent of technical change as 'the relative change in total unit costs when the techniques in each period are those which would minimize unit costs when factor prices are constant'. Similarly, 'the labour or capital-saving biases of technical advance are measured by the relative change in capital per labour unit when factor prices are constant'. Thus Fig. 2a depicts neutral technical change, that of Fig. 2b is labour-saving, and that of Fig. 2c capital-saving. This definition of innovative bias is but one of several available ones, each of which may have something to be said for it in a given context. But I believe that Salter is right in that his definition is the more appropriate one for the study of technical change at the industry level, while the others are constructed for the purpose of aggregative analysis.[5] Moreover, the industry (or firm) level is also the appropriate one for explanatory purposes, since, as explained in Ch. 3, we want to understand aggregate technical change as the outcome of individual actions. We shall see in Ch. 7 that fatal ambiguities may arise when the notion of 'labour-saving' innovations is used indiscriminately in micro-analysis and in macro-analysis.

How, then, can we explain the extent and the direction of technical change in a neoclassical theory? It must be said roundly that neoclassical theory is not well suited to this task. It is a supremely efficient tool for equilibrium analysis of economic life, including intertemporal equilibria, steady-state growth, and other phenomena that take place in logical, as opposed to historical, time. It is, on the other hand, conceptually ill at ease when dealing with genuinely dynamic problems. It remains to be seen, however, whether this is because neoclassical theory is somehow defective, or because the task of explaining the unpredictable is inherently difficult.

Neoclassical economics is committed to the explanation of all phenomena in terms of rational choice within constraints. In the present case, this implies that the rate and the bias of technical change should result from deliberate choice, presumably by the entrepreneur. But then we must ask: What are the constraints? And how can they be known to the entrepreneur? If he has access to better methods, how is it that he is not already using them? For the present I shall postpone this question, by assuming that there is an exogenously given stream of innovations available to the entrepreneur, and that his only choice is whether he shall steer them in some particular direction. I shall disregard, that is, the question of explaining the rate of change and focus exclusively on the factor-saving bias.

b. Explaining the factor-bias of technical change

The orthodox position on the issue of factor-bias was for a long time very simple: a high price of labour (respectively capital) leads to labour-saving (capital-saving) innovations. Innovation is exactly like substitution in being steered by relative prices. The classic statement of this view is found in John Hicks's *Theory of Wages* from 1932:

> The real reason for the predominance of labour-saving inventions is surely that which was hinted at in our discussion of substitution. A change in the relative prices of the factors of production is itself a spur to invention, and to invention of a particular kind – directed to economising the use of a factor which has become relatively more expensive. The general tendency to a more rapid increase of capital than labour which has marked European history during the last few centuries has naturally provided a stimulus to labour-saving invention.[6]

An equally classic rebuttal of this persuasive view was made by Salter, in his *Productivity and Technical Change* from 1960:

> If [Hicks's] theory implies that dearer labour stimulates the search for new knowledge aimed specifically at saving labour, then it is open to

serious objections. The entrepreneur is interested in reducing costs in total, not particular costs such as labour costs or capital costs. When labour costs rise any advance that reduces total cost is welcome, and whether this is achieved by saving labour or capital is irrelevant. There is no reason to assume that attention should be concentrated on labour-saving techniques, unless, because of some inherent characteristic of technology, labour-saving knowledge is easier to acquire than capital-saving knowledge.[7]

The point may also be made in slightly different terms. Assume an initial change in factor prices. The entrepreneur, then, will first substitute one factor for another until a new equilibrium has been reached, as in Fig. 1 above. In this new situation one can no longer say that one factor is more expensive than another, for in equilibrium all factors are equally expensive or equally scarce. When little is employed of one factor, then it is more productive and it is all right to pay more for it. This implies in turn that there is no further incentive to economize on the factor, since that has already been taken care of through substitution. True, the conclusion seems counter-intuitive, but I believe there is no flaw in the logic. If as a consequence of, say, dearer labour there is a trend towards labour-saving inventions, this cannot be explained along rational-choice lines. There may, however, be a causal explanation of why attention is focused on a particular kind of invention. The factor that has become relatively more expensive may come to be more salient and, by drawing attention to itself, also guide the search for new methods.

Still this is probably not the whole story. I shall shortly take up some proposed explanations of the bias in terms of rational choice, but let me first suggest that the strong intuitive appeal of the Hicksian argument may be due to an easily committed logical fallacy. Assume that the price of labour relative to capital is increasing throughout the economy, so that all entrepreneurs simultaneously face rising labour costs. It is then clear that if they all go in for labour-saving innovations, there will be a fall in the aggregate demand for labour, leading to a fall in wages. Collectively, that is, labour-saving innovations seem to be the rational response of the entrepreneurs to rising wages. But of course entrepreneurs act individually, not collectively, and so the proposed explanation fails. External economies cannot motivate behaviour under perfect competition; to believe that they can is to make a mistake closely related to the fallacy underlying functional explanation. The wage rate is a parameter for the individual entrepreneur, and he can do nothing to change it through his own behaviour. Still, if for some reason – perhaps the one suggested in the preceding paragraph – there is in fact a tendency to labour-saving innovations, these collective benefits will arise as a by-product. As in other cases, it is then very tempting to turn these by-products into reasons for the action.[8]

Observe that the situation may be understood in one of two ways. In one model we may postulate that the entrepeneur pays a cost for giving a particular bias to his search for innovations. This gives us a Prisoners' Dilemma: it is better for all entrepreneurs if all mechanize than if none does so, but for the individual entrepreneur it is tempting to defect and benefit from the labour-saving inventions undertaken by the others without making a contribution himself to the public good. And of course if the others do not mechanize, the individual entrepreneur has a disincentive to do so, since it would involve only costs and no benefits. In another model, we may postulate that there are no costs associated with steering the search for inventions in a specific direction. We then get a somewhat curious situation, which can be modelled by the following game. Using the terminology from earlier chapters, and letting 'A' stand for the mechanization strategy and 'E' for the strategy of unbiased search, we see that all are indifferent between (A,A) and (E,A) and also between (A,E) and (E,E), while all also prefer the first two outcomes to the last two. For each atomistic entrepreneur it is then a matter of strict indifference whether he gives a labour-saving bias to his search, but collectively they have a preference for mechanization. John Harsanyi suggests that in such cases the rational choice is to flip a coin between A and E.[9] My view would – tentatively – be that we should make it a constraint on the notion of individual rationality to act in a collectively rational way whenever this is not individually harmful. Or, as an alternative, we might suggest that cases like this offer scope for leadership and persuasion, if the business leaders can exploit the 'zone of indifference' of individual entrepreneurs.[10]

William Fellner[11] has suggested two mechanisms that would make it individually rational to react to dearer labour with labour-saving inventions, even assuming that there are costs associated with such bias. First, we may drop the assumption of perfect competition and assume that the firm is so large that it can internalize some of the external economies created by mechanization. If the firm, that is, to some extent acts as a monopsonist in its dealings with labour, then it can influence the wage rate by the amount of labour it hires, which in turn may be modified by mechanizing. Secondly, and more interestingly, the firm can learn from past experience. If there has been a trend of rising wages in the past, which is expected to continue into the future, then it may be rational to search for labour-saving innovations in order to preempt to some extent the impact of future wage rises. This argument works well if firms form their expectations simply on the extrapolation of past trends, but one might ask whether they do in fact do so. And, perhaps more to the point, is it rational for them to do so?

Consider again the situation as a strategic one. The individual entrepreneur may, initially, carry through the argument suggested by Fellner, but then he may come to remember that his is not the only firm in the economy. There are also other entrepreneurs in a similar situation of rising wages and, on the basis of similar reasoning, they may be expected to mechanize in order to reduce the impact of future wage increases. But if they can be expected to do this, there will be a fall in the aggregate demand for labour, and so the expected wage rate will not rise after all, which means that he has no incentive – a disincentive, in fact – to mechanize. But after further reflection he may recognize that the other entrepreneurs may be going through the same argument and reaching the same conclusion, in which case wages *will* continue to rise and mechanization again begins to look attractive. But then . . . Clearly, the situation is that of a game of Chicken, in that each entrepreneur has an incentive to mechanize if no others do so and a disincentive if all others do. This implies that it is difficult to give a full rational-choice explanation of labour-saving innovation along Fellner's lines: such innovations are rational only on the basis of expectations that are hard to defend rationally.

A different rational-choice explanation of factor-bias in technical change was proposed by Charles Kennedy.[12] He assumes that at any given time the firm confronts an *innovation possibility frontier,* which imposes constraints on the technically possible inventions. A similar model was introduced by S. Ahmad.[13] 'Kennedy suggested that the frontier should relate the *proportion* of one factor that could be saved to the *proportion* saved of the other factor. Ahmad, on the other hand, suggests that the frontier should relate the *amount* of one factor saved per unit of output to the *amount* of the other factor saved.'[14] Neither author, however, gave any reason for believing the one or the other to be the more plausible postulate. More fundamentally, neither gave any reason for believing that the innovation possibility frontier has a psychological reality for the entrepreneur. Rational-choice explanations, to insist on a point repeatedly made in Ch. 3 above, turn upon desires and beliefs: the actor chooses the action that makes most sense *to him,* not the action that in some abstract or absolute sense would be optimal. To invoke an unobservable and arguably unknowable set of feasible innovations is to take leave of this fundamental methodological principle. There may well be, at any given point in time, objective limits to the innovations that can be made on the basis of existing technical knowledge, but these limits can make no difference for behaviour and have no explanatory power unless they somehow manifest themselves to the agents. Kennedy, it seems to me, invokes maximization without a maximizer. He assumes that the innovation will occur at the point of the frontier that, at the ruling factor prices, permits the greatest reduction in unit cost, but he does not tell us how the entrepreneur is supposed to find the frontier and move along it until he finds a maximum, let alone how he is to find the global maximum. The theory, in a word, lacks microfoundations.[15]

c. Explaining the rate of technical change

The problem of explaining the rate of innovation is closely related to that of explaining the rate of innovative activity. The two problems are distinct because the product of innovative activity – new technical knowledge – has the peculiar property that 'every kind of final output is produced essentially only once'.[16] This implies that duplication is worthless, and that innovative activity cannot serve as a measure of innovation. Still the two are so closely related that a theory capable of explaining the one may also explain the other. In fact, the emphasis on microfoundations would seem to imply that the rate of innovations can *only* be explained in terms of the rate of innovative activity.

Innovation and technical change are not universal phenomena, but are restricted in time and space to a very small subset of historical societies. One way of approaching the problem of technical change is, therefore, to ask why it occurs in so negligible amounts in pre-industrial (and, with some exceptions, non-Western) societies. If action is to be explained as rational choice within a feasible set of options, then the absence of some specific action must be explained either by its absence from the feasible set, by the absence of some motivation that could rationally pick it out within that set, or by the absence of information that it is indeed within the set. In what follows I shall disregard the last issue, and stipulate that firms or entrepreneurs have some probabilistic knowledge about the feasible and profitable inventions that can be made. (Observe that this postulate is much weaker than the idea of an innovation possibility frontier, which assumes deterministic knowledge about how easy it is to make specific types of inventions.) We are then left with the choice between two explanations of lack of innovation: the absence of objects of innovation and the absence of the motivation to innovate. Perhaps it is true that in pre-industrial societies there was a pressure towards conspicuous consumption that was incompatible with the motivation to innovate.[17] But there is also the possibility that there was a lack of investment objects, i.e. that the feasible innovations did not seem economically interesting.[18]

To explore the last idea, consider the distinction between the private and the social returns to innovation. The latter, very roughly, are the benefits that accrue to society when a given innovation is universally adopted by all producers in the relevant sector. The former include the benefits that can be reaped by the innovator himself. To assume that innovations will be forthcoming when they are (in some sense) possible and socially beneficial is to ignore that there has to be an incentive for some individual to produce them and that, for reasons peculiar to the production of information, the individual and the social returns to innovation usually are vastly divergent. Once the information has been produced, it may be difficult for the innovator to prevent other entrepreneurs from taking it over and using it free of charge. If so, he will reap only the benefits that derive from his own productive use of the new technical knowledge. True, the benefits may be enhanced by his temporary monopoly during the period while his competitors are adopting the new methods, but against this he must weigh the 'penalties for taking the lead'. Usually the monopoly will not last for long, at least not long enough to justify the innovative activity which is required.

The economic and technological history from the seventeenth century onwards shows that there are four main responses to this problem.[19] First the innovator – typically a small artisan – may keep the innovation to himself, jealously guarding his professional secrets in order to hold on to his monopoly profits as long as possible. This, of course, very much limits the kind and amounts of innovations that will be forthcoming, both because the procedure excludes innovations that for purely technical reasons are hard to keep secret and because it does not permit the social returns to act as an incentive. Secondly, the State may encourage innovative activity, much as it does for the production of other kinds of public goods.[20] The historical record shows, however, that this was not an efficient method. The State was led to reward ingenious attempts to produce perpetual motion machines and similarly unproductive ventures, effectively rewarding the inventor according to his past efforts rather than to the future benefits. Thirdly, the patent system emerged as a way of extending and stabilizing the temporary monopoly of the inventor, thereby enabling him to appropriate the social returns or at least a sufficient part of them to make the innovative effort worth while. And, lastly, monopolies emerged that because of their dominant position in the industry could appropriate a substantial part of the social returns even in the absence of a patent system.

When trying to explain the rate of innovation, the most promising neoclassical approach would seem to assume that the rate of inventions is exogenously given, and that firms face the problem of how much to invest in the effort of transforming inventions into innovations. One must assume, that is, that the growth of science is not part of the economic process itself. Clearly, this is in some cases contrary to fact,[21] but these are also the cases for which explanations may be hard to find. Scientific progress is clouded in uncertainty, in the technical sense of that term, and a firm would find it extremely difficult to justify rationally the investment it might make in basic scientific research. (It would appear, however, that the life sciences are increasingly becoming exceptions to this statement.) I shall assume, moreover, that firms are able to form some justified – probabilistic – expectations about the private costs and the social returns of transforming a given invention into a productive innovation. This assumption seems to be roughly borne out by experience, although it would clearly be absurd to go for high precision in such cases.

Against this background, the standard neoclassical approach is to explain the rate of innovation in terms of the two mechanisms singled out in the paragraph preceding the last: appropriability and market structure. In addition there is obviously a problem about the demand for innovations, since the social return depends crucially on this element. Since Adam Smith it has been recognized that the amount of innovation depends on the extent of the market, a proposition that has also been confirmed in many recent studies.[22] But this aspect of the problem is somewhat incidental to the rational-choice approach favoured by neoclassical theory; and in any case I am not going to discuss it here. The final statement of the problem, then, is the following: assuming as exogenously given the scientific source of innovations and the economic demand for innovations, how do we explain the rate at which they are forthcoming?

As noted in Ch. 2 above, the paradox of the patent system is that, in dispelling dynamic suboptimality, it also destroys static optimality. By giving the innovator a temporary monopoly on innovation, it ensures that innovations will indeed be forthcoming, but also prevents them from being optimally employed. (A somewhat similar, although basically unrelated, paradox is created by the role of expectations in innovation: the more new innovations are expected to emerge in a given field, the smaller the rate at which they are diffused, since users prefer to wait for the next and improved version.)[23] From one point of view, therefore, the patent system may be included among those 'relations of production' that 'fetter' the productive forces. I discuss these Marxist notions in Appendix 2. Here I only want to stress that the patent system can be designed so as to be optimal from the social point of view, e.g. by varying the permitted fees for licensing, the duration of the patent etc.[24] The comments below on the behaviour of firms must all be understood as being relative to a given patent system, which may or may not be designed so as to channel that behaviour into socially desirable forms.

Assume, then, an industry with would-be innovating firms. How much will they rationally invest in innovative activity? And how much innovation will result? It turns out that these questions are much harder to answer than was previously thought, because they are essentially of a general-equilibrium nature. Traditionally, one assumed that the market structure was exogenous to, and provided a causal explanation for, the rate of innovative activity. Thus the 'Schumpeterian hypothesis', of which more in the next chapter, argued that monopolies tend to promote innovation. But it now turns out that 'industrial concentration and research intensity are simultaneously determined'[25] on the basis of, among other things, the internal technology of the research and development process. In a model proposed by Glenn Loury, for instance, the industry will be perfectly competitive (with an infinite number of firms) if the technology for innovation has decreasing returns to scale, but with initially increasing returns there may be an equilibrium with a finite number of firms.[26] Another failure of the partial-equilibrium analyses stems from the assumptions they make about the firm's conjecture of the behaviour of other firms. If the model, for instance, assumes that all firms are identical, one cannot determine the optimum for a given firm with respect to *given* levels of investment by other firms, since it is a constraint on the solution that in equilibrium all firms must behave identically.[27]

This also implies the constraint that in equilibrium all firms must make similar assumptions about each other. The question is whether these expectations exist, i.e. whether the 'game of innovation' has a solution. I cannot enter into the details of this discussion, if only because I feel less than confident that I have fully mastered them. Let me, however, following a model offered by Partha Dasgupta and Joseph Stiglitz, indicate why in such 'R & D games' there need not be a solution.[28] They assume that there is no uncertainty about the innovative possibilities, so that there is a known and deterministic relation between the amount invested in innovation and the time at which the innovation is forthcoming. It is then proved (i) that there must be at least one firm investing in innovation at equilibrium, (ii) that there cannot be more than one firm investing in innovation at equilibrium, and (iii) that the assumption that at equilibrium a single firm invests in innovation is inconsistent. It follows that there cannot be an equilibrium point in the game.

The reasoning behind these propositions is as follows. It is stipulated in the model that innovation is profitable, in the sense that there exists an amount x such that a firm investing x in innovation and getting the full reward will increase the present value of its profits. If in equilibrium there were two firms investing in innovation, they must invest equal amounts. But this could not be an equilibrium, for one firm could then marginally raise its investment and thus raise the chance of getting the reward from 50 per cent to 100 per cent. Finally, if in equilibrium there was just one firm investing, it would have to invest an amount that gives zero expected profits, for otherwise competitors would emerge. But since we assume this to be an equilibrium in the game-theoretic sense (a 'Nash-Cournot equilibrium', as it is usually called in the present context), the behaviour of the investing firm must be optimal with respect to that of the non-investing firms. This, however, could not be the case with zero expected profits, since on the assumption that other firms do not invest, the investing firm could obtain positive profits by reducing its R & D expenditure. It is clear that, basically, we are dealing with a variant of the game of 'Chicken': if others invest, the individual firm has no incentive to invest; if they do not, it has.

In the model, then, there is no equilibrium in pure strategies. We know, however, from game theory that there must be an equilibrium in mixed strategies, i.e. an outcome in which the firms invest stochastically in innovation. But this, of course, does not guarantee that the game has a *solution* in mixed strategies. Economists, in fact, tend to be satisfied with the demonstration that a given interaction structure has an equilibrium, i.e. an outcome that is individually stable in the sense explained earlier. Whether this equilibrium will in fact be realized, is a different matter. In a game without a solution there is no equilibrium that will be realized by the actors tacitly converging on a unique set of strategies. To assume that an equilibrium will be realized by some 'invisible hand' is, once again, to fall victim to some variety of functionalism. It could of course happen that an equilibrium would be realized by some dynamic process of mutual adjustment converging to a stable situation, but this would have to be proved in each specific case.[29]

The family of models discussed here can also throw some light on the relation between the rate of innovative activity and the rate of innovation. In Loury's simple and attractive model he first proves three theorems for the case of a fixed number of firms, with barriers to free entry: the firm's level of investment declines with an increasing number of firms; the rate of innovation increases with the number of firms; and there will be too much investment in innovation relative to the social optimum, i.e. too much innovative activity compared to the innovations that are produced. In the case of an endogenously determined number of firms (with free entry), the first two theorems also hold good, and it is also shown that with initial returns to scale in the technology of innovation there will be over-investment in innovative activity. To be sure, these results are strongly dependent on very specific assumptions; they are cited here only to give a flavour of the problems that are attacked by this sophisticated branch of neoclassical economics.

I tend to be somewhat sceptical as regards the explanatory power of these neoclassical models of technical change. The problem, basically, is that there is too much uncertainty in the situation for rational choice to be well-defined. There is, first, the uncertainty surrounding the innovative possibilities. I have assumed that this uncertainty may to some extent be reduced by learning from experience, so that the firms rather face a problem of decision under risk. This is a large concession to make, but perhaps not much larger than what is usually required in economic models. The real problem is elsewhere, in the radical uncertainty that stems from the strategic nature of the situation. The best we can do is to assume that firms act rationally on the basis of non-rational assumptions about each other, which means that we are really thrown back upon a causal explanation of their behaviour. Instead of assuming that firms act rationally on arbitrary beliefs, it might seem more parsimonious just to assume that they act arbitrarily, the entrepreneur being moved by what Keynes memorably called his 'animal spirits'. This, indeed, is one way of describing some of the models to be outlined in Ch. 6 below. By postulating that firms search randomly and then decide on the basis of satisficing rather than optimizing, one certainly comes closer to the actual entrepreneurial process.

5. Schumpeter's theory

Joseph Schumpeter (1883–1950) is the most influential single writer on technical change. He saw innovation – of which technological innovation is the main, but not the only, variety – as the engine of economic development. Moreover, he argued that innovations were also the main cause of the cyclical fluctuations which the economy undergoes in the course of that development. For him growth and cycle were indissociably linked, at least in the capitalist mode of production: to do away with the cycles one would have to eliminate the innovations that were the wellspring of growth. In addition to the explanation of these long historical trends in terms of innovation, he also offered an explanation of the innovative process itself. Here the key explanatory idea is that of the *entrepreneur* – a unique historical figure, of supernormal will and energy. Rather than gloss over the creative and unpredictable aspects of innovation, he made these into the cornerstone of this theory. Innovation is essentially a disequilibrium phenomenon – a leap into the dark, requiring abilities found only among the few. Schumpeter's eloquence in depicting the psychology of the entrepreneur is rivalled only by his elusiveness.[1] True, he is dealing with an elusive phenomenon, the psychology of creation. Yet this does not excuse the self-indulgent verbosity that characterizes many of his writings. Lionel Robbins is reported to have characterized one of his major works, *Capitalism, Socialism and Democracy*, as 'supremely intelligent after-dinner talk',[2] and any reader can attest to the core of truth in this assertion.

On the other hand, Schumpeter was 'the great character reader of the capitalist system', to employ a phrase used by W. Leontief to characterize Marx.[3] His vast historical learning, his intimate knowledge of the workings of the capitalist system, his realism – bordering on the cynical – and his psychological acumen made him uniquely suited to attempt a large-scale analysis of capitalist development, rivalled only by those of

Marx and Weber.[4] Like them he offered not only an analysis of capitalism as an economic system, but also an account of the political organization most congenial to it. His theory of the political entrepreneur – a direct outgrowth of the theory of economic innovation – has given rise to a whole school of political scientists, with Downs's *An Economic Theory of Democracy* as the best-known work. Economic as well as political life, in his vision, are shaped by men with the ability to 'get things done', to overcome habitual thinking and perceive objective possibilities hidden to others. Although routine activities may be predominant in the quantitative sense, innovations are the sparks that ignite and vivify the system. Hence in his work on business cycles Schumpeter was comparatively uninterested in the mechanisms of diffusion and propagation that determine the precise shape of the fluctuations, and much more concerned with the innovations that provide the irregular series of shocks to be propagated.[5]

The last observation also throws light on his deep hostility to Keynes, and his refusal to entertain the element of effective demand as a major explanation of the business cycle.[6] He felt that this could only throw light upon the superficial aspects of the process and distract attention from the basic causes. Moreover, he was strongly averse to the aggregate method used by Keynes and his followers: 'It keeps analysis on the surface of things and prevents it from penetrating into the industrial processes below, which are what really matters. It invites a mechanistic and formalistic treatment of a few isolated contour lines and attributes to aggregates a life of their own and a causal significance that they do not possess.'[7] Schumpeter not only practised the doctrine of methodological individualism: according to Fritz Machlup it was he who coined the term in the first place, and made the often-neglected distinction between political and methodological individualism.[8]

I shall treat Schumpeter's central works in chronological order, yet not hesitate to quote from one work in order to elaborate points made in others. The unity of Schumpeter's work from 1911 onwards is such that this procedure seems perfectly legitimate.

a. The Theory of Capitalist Development (1911)

Virtually all of Schumpeter's economic theory is contained in this early work (here to be quoted after the somewhat revised English translation from 1934). I shall focus, however, on that part of it which deals with the micro-economics – or rather the psychology – of innovation. Although he also discusses the macro-consequences of innovation, these are more fully treated in later works.

The basic form of innovation is qualitative and discontinuous: 'It is that kind of change arising from within the system *which so displaces its equilibrium point that the new one cannot be reached from the old one by infinitesimal steps*. Add successively as many mail coaches as you please, you will never get a railway thereby.'[9] It is as the theorist of innovation in this sense that Schumpeter is known, and it is in this capacity I shall mainly discuss him here. Yet Schumpeter also has suggestive ideas – the term 'theory' would be too pretentious – about technical change occurring through infinitesimally small steps. It will prove useful for several purposes to quote and briefly discuss his account of such 'small' and 'adaptive' – as opposed to 'large' and 'spontaneous' – improvements in technique:

> The assumption that conduct is prompt and rational is in all cases a fiction. But it proves to be sufficiently near to reality, if things have time to hammer logic into men. Where this has happened, and within the limits of which it has happened, one may rest content with this fiction and build theories upon it. It is then not true that habit or custom or non-economic ways of thinking cause a hopeless difference between the individuals of different classes, times or cultures, and that, for example, the 'economics of the stock exchange' would be inapplicable, say, to the peasants of to-day or to the craftsmen of the Middle Ages. On the contrary the same theoretical picture in its broadest contour lines fits the individuals of quite different cultures, whatever their degree of intelligence and of economic rationality, and we can depend upon it that the peasant sells his calf just as cunningly and egotistically as the stock exchange member his portfolio of shares. But this holds good only where precedents without number have formed through decades and, in fundamentals, through hundreds and thousands of years, and have eliminated unadapted behaviour.[10]

Schumpeter adds that such 'small variations at the margins' may 'in time add up to great amounts'.[11] Given this, it is not clear why the adaptive technical change should be less important than the spontaneous variety.

Clearly, further argument is needed. Schumpeter does not provide this, but we shall have occasion to return to the problem, in this chapter and in the following.

'Adaptive technical change' permits functional explanation of economic behaviour. As Schumpeter observes, 'There may be rational *conduct* even in the absence of rational *motive*.'[12] The text is not sufficiently precise to allow us to decide whether Schumpeter had in mind some fairly direct analogy of natural selection, or something more similar to the Markov-processes considered in Ch. 2 above. But we may note that Schumpeter insists on the great time span required for such adaptation to work itself out. It follows that in a rapidly changing environment there will not be time enough for 'rational conduct' to become fixed, and even should this happen, it would soon cease to be rational. Incremental adaptation – to anticipate – can at most achieve local maxima, and even these will be unattainable in a rapidly changing environment.

Mark Elvin has recently discussed Chinese economic history in terms quite similar to Schumpeter's adaptive change, although Schumpeter is never cited in his work.[13] Elvin's central problem is why the Chinese economy, after a phase of vigorous growth, was caught in a 'high-level equilibrium trap' from which it could not escape by the traditional incremental productivity increases. The puzzle was well stated by R. H. Tawney: why is it that 'China ploughed with iron when Europe used wood, and continued to plough with it when Europe used steel'?[14] Elvin points out that China eventually reached a point at which only discontinuous technical change could have brought about further productivity increases, since the potential of the traditional technology had already been used up. Why, then, did China not make that leap? Not because Chinese science was undeveloped: 'In most fields, agriculture being the chief exception, Chinese technology stopped progressing well before the point at which a lack of basic scientific knowledge had become a serious obstacle.'[15] Nor because innovation was foreign to the Chinese mentality: this was a civilization 'with a strong sense of economic rationality, with an appreciation of invention such that shrines were erected to historic inventors (though, it is true, no patent law), and with notable mechanical gifts.'[16]

In Elvin's own, complex, explanation of the technological stagnation in China, a key passage is the following: 'Huge but nearly static markets created no bottlenecks in the production system that might have

116 THEORIES OF TECHNICAL CHANGE

prompted creativity. When temporary shortages arose, mercantile versatility, based on cheap transport, was a faster and surer remedy than the contrivance of machines.'[17] Against this, a neoclassical economic historian would stress the lack of a patent system (mentioned only parenthetically by Elvin) or of some similar arrangement for internalizing the social benefits of innovation.[18] And Schumpeter, while also emphasizing the need for some such arrangement, would add that the stagnation of the Chinese economy is linked to the absence in China of the *innovating entrepreneur*.

Capitalist innovation, for Schumpeter, is a much broader concept than the idea of technical innovation carried out by a capitalist firm. Innovation is defined quite generally as *the carrying out of new combinations of the means of production*, and includes the following cases:

(1) The introduction of a new good – that is one with which consumers are not yet familiar – or of a new quality of a good. (2) The introduction of a new method of production, that is one not yet tested by experience in the branch of manufacture concerned, which need by no means be founded upon a discovery scientifically new, and can also exist in a new way of handling a commodity commercially. (3) The opening of a new market, that is a market into which the particular branch of manufacture of the country in question has not previously entered, whether or not this market has existed before. (4) The conquest of a new source of supply of raw materials or half-manufactured goods, again irrespective of whether this source already exists or whether it first has to be created. (5) The carrying out of a new organisation of any industry, like the creation of a monopoly position (for example through trustification) or the breaking up of a monopoly position.[19]

This defines innovation and the entrepreneurial function generally, but Schumpeter is concerned only with capitalist innovation. In *Business Cycles* Schumpeter offered the following definition of capitalism: it is 'that form of private property economy in which innovations are carried out by means of borrowed money, which in general, though not by logical necessity, implies credit creation'.[20] This, for instance, enables him to suggest that the German principalities in the eighteenth century carried out capitalist innovation, by public servants financing their enterprises commercially.[21] It also allowed him to make a famous and dubious claim that entrepreneurial profits are in no way a return for risk-taking, since the financial risks are assumed by the capital lender.[22]

Let me focus on innovation in the narrow technical sense, and ask how Schumpeter explains the extent and the timing of entrepreneurial innovation. (Since Schumpeter never raised the issue of the factor-bias of technical change, this problem falls outside the scope of the discussion.)[23] What makes the entrepreneur tick? Is innovation a rational activity? We have seen that the neoclassical answer to these questions is that the entrepreneur is out to maximize profits, and that he innovates to the extent that innovation is a rational means to this end, given more or less rational assumptions about the costs and benefits of innovation and about the behaviour of other entrepreneurs. Schumpeter has somewhat different answers to both questions. In his discussion of entrepreneurial motives he emphasizes three elements: the dream and the will to found a private kingdom; the will to conquer, to succeed for the sake not of the fruits of success, but of success itself; and finally the joy of creating, of getting things done.[24] Only the first of these is directly linked to the acquisition of private property, although pecuniary gain is also a very accurate indicator of the extent to which the other desires are realized.[25]

When we look more closely at these entrepreneurial motives, as depicted by Schumpeter, they appear remarkably elusive. First, it is not clear whether the entrepreneur is basically rational or guided by more 'atavistic' motives, to use a phrase employed by Schumpeter in his study of imperialism.[26] The 'will to conquer' is both limitless and objectless. It has no precisely defined object that could turn it into a rational pursuit, but feeds upon itself in an indefinite quest for accumulation. 'The modern businessman acquires work habits because of the need for making a living, but labors far beyond the limits where acquisition still has a rational meaning in the hedonistic sense.'[27] The analogy with Marx is clear enough, despite claims to the contrary.[28] Such arguments, then, seem to justify the view that 'In a master stroke of postulational economy, Schumpeter was able to bring in his "psychology of the entrepreneur" as a special case of his principle of "overlapping *geists*".'[29] On the other hand, however, Schumpeter stressed the 'rationalist and unheroic' character of the bourgeois.[30] This, indeed, is such a pervasive theme of his work that it is hard to see how it can be reconciled with the notion of the atavistic dynasty founder who wants success rather than the fruits of success.

The preceding comments addressed themselves to the problem of entrepreneurial motives and desires. Similar comments apply to Schumpeter's theory of how entrepreneurial beliefs and expectations are formed. In *Business Cycles* he manages to say within the scope of a single paragraph both that businessmen invariably are too optimistic and that they are far from always being so.[31] More central, however, is the following ambiguity. In *The Theory of Economic Development* he seems to argue that entrepreneurs are invariably right in their evaluation of innovative possibilities: their success 'depends on intuition, the capacity of seeing things in a way, which afterwards proves to be true, even though it cannot be established at the moment, and of grasping the essential fact, discarding the unessential, even though one can give no account of the principles by which this is done.'[32] Yet in *Capitalism, Socialism and Democracy* we meet an argument apparently contradicting this, in a passage that I shall quote at some length since it conveys well the characteristic virtues and vices of Schumpeter's style:

Bourgeois society has been cast in a purely economic mold: its foundations, beams and beacons are all made of economic material. The building faces toward the economic side of life. Prizes and penalties are measured in pecuniary terms. Going up and going down means making and losing money. This, of course, nobody can deny. But I wish to add that, within its own frame, that social arrangement is, or at all events was, singularly effective. In part it appeals to, and in part it creates, a scheme of motives that is unsurpassed in simplicity and force. The promises of wealth and the threats of destitution that it holds out, it redeems with ruthless promptitude. Wherever the bourgeois way of life asserts itself sufficiently to dim the beacons of other social worlds, these promises are strong enough to attract the large majority of supernormal brains and to identify success with business success. They are not proffered at random; yet there is a sufficiently enticing admixture of chance; the game is not like roulette, it is more like poker. They are addressed to ability, energy and supernormal capacity for work; but if there were a way of measuring either that ability in general or the personal achievement that goes into any particular success, the premiums actually paid out would probably not be found proportional to either. Spectacular prizes much greater than would have been necessary to call forth the particular efforts are thrown to a small majority of winners, thus propelling more efficaciously than a more equal and more 'just' distribution would, the activity of that large majority of businessmen who receive in turn very modest compensation or nothing or less than nothing, and yet do their utmost because they have the big prizes before their eyes and *overrate their chances of doing equally well.*[33]

Here the key passage that I have italicized states that the capitalist system works so well because it induces unrealistic expectations about success and thereby draws out much more effort than would have been forthcoming from more sober spirits. This theme – the social benefits of individual self-deception – has been raised in a number of more recent contributions as well, from Albert Hirschman's 'Principle of the Hiding Hand'[34] to studies in cognitive psychology. Schumpeter's argument, although elusive, may be related to the following:

People sometimes may require overly optimistic or overly pessimistic subjective probabilities to goad them into effective action or prevent them from taking dangerous action. Thus it is far from clear that a bride or groom would be well advised to believe, on their wedding day, that the probability of their divorce is a high as .40. A baseball player with a batting average of .200 may not be best served, as he steps up to bat, by the belief that the probability that he will get a hit is only .2. The social benefits of individually erroneous subjective probabilities may be great even when the individual pays a high price for the error. We probably would have few novelists, actors or scientists if all potential aspirants to these careers took action based on a normatively justifiable probability of success. We also might have few new products, new medical procedures, new political movements, or new scientific theories.[35]

The first half of the passage suggests that excessive optimism may benefit the individual: in order to achieve anything at all, he does well to believe that he will achieve much. In other cases, as suggested by the second half, only society gains. The excessively sanguine expectations ensure that a large number of individuals enter the competition, so that the material on which selection – by the market or some other screening mechanism – can act is as large as possible and the selected individual as good as possible. The losers, then, contribute nothing: they are only the incidental byproduct of the process of selection. The winners, by contrast, are those whose expectations prove, *ex post*, not to have been too sanguine: they are the infallible individuals referred to by Schumpeter in the passage quoted earlier. An alternative interpretation of Schumpeter would be that even the losers have a social function, since they are induced by the lure of reward to work harder than they would otherwise have done. This, presumably, would be the case with science: most people who choose a scientific career do so because they wrongly believe they will be outstanding, and yet they do offer a positive contribution beyond that of providing material for selecting those who will prove outstanding, since routine scientific work is also needed for progress to occur.[36]

To conclude, then, it is hard to decide whether Schumpeter conceives the motives and expectations behind innovation as rational or irrational. In a sense, the innovating entrepreneur exhibits 'rational conduct without a rational motive', since he has expectations that *ex ante* are irrational and only *ex post* are confirmed. His behaviour is rational in the sense of successfully exploiting the objective possibilities of innovation, yet irrational in that he is ridden by a demon who never lets him be satisfied by results. At any rate it would not make sense to describe his behaviour as the choice of the best member of a known feasible set. His gift is in expanding the feasible set, not in choosing rationally within it.

b. Business Cycles (1939)

In this monumental work of more than 1000 pages Schumpeter studied the macro-consequences of innovation – economic growth and the business fluctuations that invariably accompany it. This theory of business cycles has proved difficult to model mathematically, and for very good reasons. Both R.M. Goodwin and Ragnar Frisch have constructed semi-Schumpeterian models of the business cycle in which a steady increase in innovative possibilities is transformed into discontinuous spurts of actual innovations.[37] On Goodwin's model, for example, the innovative potential is automatically transformed into innovation once it exceeds a certain critical level. This, however, is a betrayal of Schumpeter's thought. As noted by A.P. Usher, on Schumpeter's view innovations 'are neither transcendental and unknowable, nor mechanical and foreordained'.[38] In particular, innovations are not responses to pre-existing needs, since they often create the very need they satisfy. Schumpeter cites the motor car as an example.[39] More important than the size of the potential profit is the presence of an entrepreneur capable of 'doing the thing'. Schumpeter cites the cotton and wool industries in late eighteenth-century England as examples: innovative possibilities were present in both, but only the cotton industry had the New Men capable of carrying out the innovations.[40] In other words, old industries fail to innovate not because of diminishing opportunities of innovation, but because of a weakening of the motivation to innovate.[41] Moreover, when capitalism arrives at a stage in which objective possibilities for innovation are automatically transformed into innovation, the age of the entrepreneur has passed.[42] Schumpeter's theory, then, is like the theory of natural selection in allowing only for explanation of what actually happened, not for prediction of what will happen. Broadly speaking, the innovations are unpredictable in the same sense that mutations are unpredictable. The course of history is not shaped by the aggregate mass of innovations, but by outstanding individual innovations that depend on the random appearance of exceptionally gifted individuals. There are, in other words, no 'forces' in history that channel individual efforts into specific directions and make for homeorhesis in the sense explained in Ch. 1 above.

Business Cycles is, among other things, a running argument for the following two propositions about capitalism. (i) In capitalism the phenomenon of economic growth is indissociably linked with cyclical fluctuations. (ii) In capitalism, the long-term efficiency is obtained only at the price of short-term inefficiency. These are different propositions, since the last does not by itself imply anything about cycles. I shall discuss them in the order indicated.

Schumpeter clearly dissociates himself from what he calls the 'Marshall-Moore theory of organic growth', according to which economic development exhibits 'first a smooth and steady movement, and second fluctuations around it which are due to random shocks':[43] a cycle superimposed on an independent trend. The engine of growth on this conception could either be savings or widening markets; in any case the innovative entrepreneur plays no role. Against this, Schumpeter embraces what can only be called a Hegelian theory of economic progress:

We must cease to think of [progress] as by nature smooth and harmonious in the sense that rough passages and disharmonies present phenomena foreign to its mechanism and require special explanations by facts not embodied in its pure model. On the contrary, we must recognize that evolution is lopsided, discontinuous, disharmonious by nature – that the disharmony is inherent in the very *modus operandi* of the factors of progress. Surely, this is not out of keeping with observation: the history of capitalism is studded with violent bursts and catastrophes which do not accord well with the alternative hypothesis we herewith discard, and the reader may well find that we have taken unnecessary trouble to come to the conclusion that evolution is a disturbance of existing structures and more like a series of explosions than a gentle, though incessant, transformation.[44]

The mechanism by which innovation makes for disturbance and disharmony is fairly complex. Schumpeter distinguishes between recession and depression. The first comes about because the introduction of new methods forces other firms to adapt and rationalize or become extinct; also because they create a situation of disequilibrium in which rational calculation becomes impossible. These are pains of adjustment that by themselves to not imply a cyclical movement, i.e. there is no reason why the recession at any stage should dip below the pre-innovative level. In fact, an aggregate index of output 'will display nothing except increase. But mere increase in total output would not produce [the effects of disequilibrium]. It is disharmonious or one-sided increase and shifts *within* the aggregative quantity which matter.'[45] Depression, however, belongs to what Schumpeter calls 'the secondary approximation', and is less fundamental than recession. It comes about, basically, because of irrational expectations, e.g. because 'many people will act on the assumption that the rates of change they observe will continue indefinitely'.[46] Schumpeter, in other words, here invokes the collectively disastrous results that may emerge when individuals act on parametric rationality. The argument today sounds obvious and unexciting, and may even have been so at the time of writing. It is then more than curious to see that Schumpeter dismisses Tinbergen's model of the shipbuilding cycle by observing that it 'violates not only the general assumptions of economic theory about rationality of behavior but is at variance also with common sense. We cannot reasonably assume that reaction to, say, abnormally favorable freight rates will be mechanical and proceed without any considerations of the causes and the probable duration of that state of things and of the effects of simultaneous action by the whole trade'.[47] Here Schumpeter assumes as a matter of course that people will form strategic or rational expectations.

Recession and depression differ in several respects in Schumpeter's scheme. Recession has a task to perform, the travail of adjustment, and ends when the task is accomplished. Depression, by contrast, 'has a way of feeding upon itself'[48] and may go into a vicious spiral. For this reason depression will actually bring about a cycle, i.e. a plunge below the pre-innovative equilibrium. Also, 'the case for government action in depression . . . remains . . . incomparably stronger than it is in recession'.[49] We may note in this connection that the semi-Schumpeterian analysis by Nicholas Kaldor[50] tends to invert the causal link sketched above, by arguing that growth is caused by irrational and excessively optimistic expectations, rather than such expectations creating a 'pathological process to which no organic functions can be attributed'.[51] For Schumpeter, the cycle is strictly a by-product of growth, whereas for Kaldor it is causally efficacious in inducing growth. (Cf. also the distinction made in Ch. 2 between the two varieties of the seventeenth-century theodicy.)

The 'Schumpeterian hypothesis' or the 'Schumpeterian trade-off' referred to in Ch. 4 states that innovations are favoured by oligopoly. The patent system is but one of the many oligopolistic or monopolistic practices which favours innovation at the expense of static allocative efficency. A recent statement is the following: 'Where patent protection is spotty and imitation may occur rapidly, the payoff to an innovator may depend largely on his ability to exploit that innovation over a relatively short period of time. Large firms have a level of production, productive capacity, marketing arrangements, and finance that enables them quickly to exploit a new technology at relatively large scale.'[52] The empirical plausibility of the hypothesis may not seem compelling. A recent survey distinguished between two senses of the independent[53] variable: firm size and market structure. Similarly a distinction was made between two senses of the dependent variable: amount of innovative activity and amount of innovation. Of the ensuing four versions of the hypothesis none was unambiguously confirmed by the evidence.[54] I shall leave these empirical matters aside, however. The Schumpeterian hypothesis may well throw light on the history of capitalism even if it does not compare well with data from the last half-century. In any case, the very existence and uncontroversial importance of the patent system shows that there is an unassailable core of truth in the hypothesis, even should it turn out to be the case that other forms of oligopolistic practices do not have similar long-run benefits. The best-known statement of the hypothesis is found in *Capitalism, Socialism and Democracy:*

> [Since] we are dealing with a process whose every element takes considerable time in revealing its true features and ultimate effects, there is no point in appraising the performance of that process *ex visu* of a given point of time; we must judge its performance as it unfolds through decades or centuries. A system – any system, economic or other – that at *every* given point of time fully utilizes its possibilities to the best advantage may yet in the long run be inferior to a system that does so at *no* given point of time, because the latter's failure to do so may be a condition for the level or speed of long-run performance.[55]

This is what I referred to in Ch. 1 above as Tocqueville's *principle of the long term.* Going beyond Tocqueville to Leibniz, we find the same principle stated with unsurpassable precision: 'The infinite series of all things may be the best of all possible series, although what exists in the universe at each particular instant is not the best possible'.[56] The idea is capable of many applications. Schumpeter himself refers to 'the fact that producing at minimum cost over time may, in the case of lumpy factors, imply never producing at minimum cost on any of the short-time cost curves because another "method" becomes advantageous before that point is reached'.[57] Kenneth Arrow argues that the Schumpeterian entrepreneur himself embodies a similar dilemma: by maximizing utility rather than profits, he prevents the realization of efficient states, but at the same time it is his personal ties with the firm (and the subsequent maximization of utility rather than profits) which makes him into the Prometheus of growth.[58] Similarly, it seems plausible that many attempts to justify the 'backyard steel furnaces' and similar practices of the Great Leap Forward in China had this general structure. True, it was said, the mass line implies a loss of efficiency, but this is more than made up for by the increase in human effort and energy that it unleashes. Or again, participation in industry is justified by the idea that there is a positive net effect of two opposing tendencies, viz. loss of efficiency in decision-making and a gain in motivation.

One should add, however, that Schumpeter did not insist on the idea that oligopolistic practices had negative consequences for the static allocation of resources. He deliberately went against the current, and argued that in many or even most cases oligopolies will cut prices and expand production rather than follow the textbook argument.[59] In fact, if a firm tried to exploit a monopolistic position it had somehow acquired, it would soon lose it.[60] Schumpeter's theory, therefore, is two-pronged:

> [Imperfections of either competition or equilibrium] yield so strong and so universally recognized a case against the pretended efficiency of the capitalist machine that it becomes necessary to recall on the one hand that a system in which imperfect competition prevails will, contrary to established opinion, produce in very many, perhaps in most cases, results similar to those which could be expected from perfect competition and, on the other hand, that even if a system consistently turned out less than its optimum quantity, this would not in itself constitute disproof of optimal performance over time.[61]

Of these two arguments, posterity has retained only the second, while arguably the first was to Schumpeter the more important one, at least in *Business Cycles*.

c. Capitalism, Socialism and Democracy (1942)

Of the many themes struck in this many-faceted and endlessly fascinating book I shall only mention two: the idea that capitalism will be destroyed by its own success, and the theory of democracy as competition for votes. And I shall not discuss these with the thoroughness they deserve, but only in so far as they are connected with the main themes of the present book.

From the general definition of innovation given in *The Theory of Capitalist Development* it follows that the creation of new organizational forms is one of the tasks performed by the innovating entrepreneur. In particular, the creation of the giant concern characteristic of 'Trustified Capitalism'[62] requires an act of entrepreneurship – while also setting the stage for the elimination of the entrepreneur.[63] This is so, Schumpeter argued, because the giant concern turns innovation into routine, so that the 'objective possibilities' for innovation are automatically carried into effect.[64] The broad social and political consequences of the disappearance of the entrepreneur are sketched in a famous passage:

> If capitalist evolution – 'progress' – either ceases or becomes completely automatic, the economic basis of the industrial bourgeoisie will be reduced eventually to wages such as are paid for current administrative work excepting remnants of quasi-rents and monopolized gains that may be expected to linger on for some time. Since capitalist enterprise, by its very achievement, tends to automatize progress, we conclude that it tends to make itself superfluous – to break to pieces under the pressure of its own success. The perfectly bureaucratized giant industrial unit not only ousts the small or medium-sized firm and 'expropriates' its owners, but in the end it also ousts the entrepreneur and expropriates the bourgeoisie as a class which in the process stands to lose not only its income but also what is infinitely more important, its function. The true pacemakers of socialism were not the intellectuals or agitators who preached it but the Vanderbilts, Carnegies and Rockefellers.[65]

The basic defect of this argument, in my view, is the idea that innovation will be routinized. At one point Schumpeter employs a more appropriate phrase, when he says that 'technological research becomes increasingly mechanized and organized'.[66] Innovation, in other words, is turned into an industry – which presumably has a production function that can be shaped and changed by innovation at the meta-level. The task of entrepreneurship will then not be to create new methods or products, but to organize research teams for the purpose of such creation. On the basis of impressionistic evidence it does indeed seem that in industries such as computers, chemicals, and biological engineering, the entrepreneurial challenge is to bring together economic, technological and scientific expertise for the overriding purpose of profit maximizing (or satisficing). Moreover, the other innovative tasks mentioned by Schumpeter, such as the opening up of new markets and the conquest of a new source of supply, will presumably not be subject to routinization in the same sense as technological innovation can be. And finally, in the interstices between the giant firms there seems to be a permanent place for small innovating firms operating at the frontiers of knowledge. For all these reasons it seems difficult to accept Schumpeter's view – clearly inspired by similar thoughts in Marx – that capitalism will be the victim of its own successes.

We have seen that Schumpeter broadened the notions of innovation and entrepreneurship so as to include not only technical change, but also institutional and organizational change. In the definition quoted above innovation was, however, limited to the economic sphere. It is, in all cases, a means to profit. In *Capitalism, Socialism and Democracy* Schumpeter took the further step of broadening the notion of entrepreneurship so as to include the political sphere as well, with power substituted for profits. True, Schumpeter never uses the term 'entrepreneur' of the political leader, nor the word 'innovation' about what he does to acquire and retain power. Yet it seems perfectly obvious that his political and his economic theories come from the same matrix.

Schumpeter propounds as a basic premise that democracy is a method for making political decisions, not an end in itself.[67] What characterizes democracy, and distinguishes it from other such methods, is 'by whom and how' these decisions are made. His answer is that democracy is 'that institutional arrangement for arriving at political decisions in which individuals acquire the power to decide by means of a competitive struggle for the people's vote'.[68] These individuals are the political leaders, the analogy of the entrepreneur in the political sphere. As in the case of the entrepreneur *stricto sensu,* they must have 'considerable personal force': they must be able to sway the electorate, mobilize followers, control rivals, subdue the opposition and the like. By contrast, it is not required that they should have any actual ability to govern. Unlike competitive capitalism – and this is an important disanalogy between the economic and the political sphere – political competition does not select for the ability to perform, but for the ability to persuade. 'The qualities of intellect and character that make a good candidate are not necessarily those that make a good administrator.'[69] From the point of view of performance in office, the selection process has at most some negative merits: 'there are after all many rocks in the stream that carries politicians to national office which are not entirely ineffective in barring the progress of the moron or the windbag'.[70] Moreover, not only are the talents of the successful politician not those of the good administrator: he will also be motivated by different concerns. The policy 'that a government decides on with an eye to its political chances is not necessarily the

128   THEORIES OF TECHNICAL CHANGE

one that will produce the results most satisfactory to the nation'.[71] Politicians seek power and then re-election, not the good of the nation.

How, then, can one characterize democracy as a method for making political decisions? Schumpeter's answer is that decisions are taken as a means to, or a by-product of, the power struggle. The following passage makes the point in eloquent terms:

> When two armies operate against each other, their individual moves are always centered upon particular objects that are determined by their strategical or tactical situations. They may contend for a particular stretch of country or for a particular hill. But the desirability of conquering that stretch or hill must be derived from the strategical or tactical purpose, which is to beat the enemy. It would be obviously absurd to derive it from any extra-military properties the stretch or hill may have. Similarly, the first and foremost aim of each political party is to prevail over the others in order to get into power or to stay in it. Like the conquest of the stretch of country or the hill, the decision of the political issues is, from the standpoint of the parliamentarian, not the end but only the material of parliamentary activity.[72]

The analogy is useful up to a point. Beyond that point – and Schumpeter definitely goes beyond it – it turns into an *idée fixe*. As in the case of war, the ultimate purpose of politics is not merely to gain a commanding position, but to exploit it. The idea that 'the department store cannot be defined in terms of its brands and a party cannot be defined in terms of its principles'[73] belongs, quite definitely, to the sphere of after-dinner talk. Salutary as a counterweight to cant, it cannot stand as a balanced view on its own. It does not fall within the scope of the present exposition to pursue this matter. Let me turn instead to the important methodological point made by Schumpeter in the following passage:

> In observing human societies we do not as a rule find it difficult to specify, at least in a rough common-sense manner, the various ends that the societies under study struggle to attain. The ends may be said to provide the rationale or meaning of corresponding individual activities. But it does not follow that the social meaning of a type of activity will necessarily provide the motive power, hence the explanation of the latter. If it does not, a theory that contents itself with an

analysis of the social end or need to be served cannot be accepted as an adequate account of the activities that serve it. For instance, the reason why there is such a thing as economic activity is of course that people want to eat, to clothe themselves and so on. To provide the means to satisfy those wants is the social end or meaning of production. Nevertheless we all agree that this proposition would make a most unrealistic starting point for a theory of economic activity in commercial society and that we shall do much better if we start from propositions about profit. Similarly, the social meaning or function of parliamentary activity is no doubt to turn out legislation and, in part, administrative measures. But in order to understand how democratic politics serve this social end, we must start from the competitive struggle for power and office and realize that the social function is fulfilled, as it were, incidentally – in the same sense as production is incidental to the making of profits.[74]

The argument is characteristically ambiguous. Schumpeter first states that only motives can explain an activity, but then goes on to say that the social end or meaning is 'the reason why there is such a thing as economic activity', as if providing a reason was not to provide an explanation. Moreover, in the economic case it would seem that the social meaning also provides an explanation in terms of motives, viz. by offering a *filter-explanation* of economic behaviour.[75] The consumers select among the would-be entrepreneurs those who in fact satisfy their wants. In the political case, however, this is not true: 'the will of the people is the product and not the motive power of the political process'.[76] The successful political entrepreneur is not one who is good at interpreting and aggregating the (pre-existing) desires of the voters, but one who is good at shaping them.

Whereas later developments of the Schumpeterian argument have assumed voters to act in a rationally self-interested way on pre-existing desires,[77] Schumpeter explicitly rejects this idea and thereby appears to '[abandon] the market economy analogy'.[78] And on reading *Capitalism, Socialism and Democracy* one might indeed get this impression. But in *Business Cycles* Schumpeter makes it clear that he has a non-standard view of the market economy, which re-establishes the analogy. He here assumes throughout that 'consumers' initiative in changing their tastes – *i.e.* in changing that set of our data which general theory comprises in the concepts of "utility functions" or "indifference varieties" – is negligible and that all change in consumers' tastes is incidental to, and brought about by, producers' action'.[79] He then, however, qualifies this stark assertion by saying that once a new type of product has been introduced, often against initial consumer resistance, the consumers will develop quality standards for the product that are not similarly manipulable. And here there is a real disanalogy between economics and politics, for in the latter domain there is not the 'pungent sense of reality' or the 'salutary and rationalizing influence of favourable and unfavourable experience'.[80] In politics, the citizen is rarely confronted with the unfavourable consequences of his choices, because the social causality is too opaque for the link between cause and effect to be perceived.[81] The entrepreneur in a real sense is constrained by the consumers: he can manipulate their desires, but not their standards of rationality and quality.[82] The politician, by contrast, meets no resistance among the citizens, provided he is sufficiently eloquent and forceful. For this reason the 'social functions' fulfilled by the entrepreneur and the politician cannot – even on Schumpeter's own assumptions – be assimilated to the extent he does in the passage quoted above.

From this discussion of Schumpeter, then, we may retain the emphasis on disequilibrium; the view that innovation occurs discontinuously; the insistence on the *embodied* aspect of innovation, i.e. on the innovating entrepreneur; the view that innovation occurs in a random – or at least far from automatic – manner; the idea that technical change requires oligopoly and possibly involves static inefficiency; the subtle discussion of the desires and beliefs of the innovating entrepreneur; and finally the notion that technical change is to be understood as a case of innovation more generally and not just as another piece of routine economic behaviour. In the discussion of the work of Richard Nelson and Sidney Winter in the next chapter, we shall see how many, but not all, of these elements have been taken over in a more precise – and inevitably somewhat different – form.

6. Evolutionary theories

The theory of evolution by chance variation and natural selection is immensely attractive, being simple as well as powerful. It is not at all surprising that social scientists should try to harness it to their explanatory purposes, be it in the form of functional explanation or in some other version. In this chapter I shall survey some evolutionary theories of technical change, beginning with the biological theory of tool behaviour and going on to the more specifically sociological theories.

a. Animal tool behaviour

In the following I draw heavily upon two valuable recent works, Benjamin Beck's *Animal Tool Behavior* and Robert Fagen's *Animal Play Behavior*. The latter has a concluding chapter on 'Play, innovation, invention and tradition' that places tool behaviour within the more general context of innovation, and offers a few models to explain how innovation could emerge – or be prevented from emerging – by evolutionary mechanisms (including cultural transmission). The former is an exhaustive catalogue and acute conceptual analysis of tool use and manufacture. Without these works of synthesis it would have been impossible for a non-specialist to grasp the problems involved. Even with their help I may have been led through ignorance into error, for which I apologise in advance.

According to Beck, there are 21 modes of tool use among animals: unaimed throwing, dragging, slapping, rolling or kicking, aimed throwing, dropping or throwing down, brandishing or waving, clubbing or beating, prodding or jabbing, reaching, pounding or hammering, inserting and probing, prying or applying leverage, digging, balancing and climbing, propping and climbing, stacking, hanging and swinging, absorbing or sponging, wiping, draping or affixing, containing, baiting.[1] These in turn may be grouped into four functional classes, according to whether they serve to 'extend the user's reach, amplify the mechanical force that the user can exert on the environment, enhance the effectiveness of the user's display behaviors, or increase the efficiency with which the user can control fluids'.[2] As for tool manufacture, Beck distinguishes four modes: detach, subtract, add or combine, reshape.[3] Many cases of tool manufacture involve several of these modes, as when 'birds first detach fibers from bits of bark and then combine the fibers to form small ovoid pellets'[4] or when 'monkeys first detach the leaves and then reshape them to increase absorbency'.[5] Beck, then, does not accept the definition of man as a tool-making animal, since this behaviour is also found among other species. He does concede that man may be unique in using tools to make tools,[6] just as among animals man is unique in the capacity to make statements about statements.[7] But he also adds that 'we have seen in the last few decades a steady erosion of putative hallmarks of man: it would not be surprising to find that some animals can use a tool to make a tool'.[8]

Beck's concern is with tools, not with innovation and technical change more generally. This means, for example, that animals pounding a shell against a stone to open it do not fall within his analysis, whereas those who pound a stone against the shell do. The reason for excluding the former is merely one of convenience. Beck does not believe that tool behaviour – as he defines it – is in any way privileged from the cognitive point of view. True, such behaviour displays a number of important cognitive features to be enumerated below, but it is by no means unique in doing so. Indeed, Beck offers an extensive analysis of gulls which capture molluscs and drop them on hard surfaces to get access to the edible interior, in order to show that this (non-tool) behaviour exhibits all the cognitively interesting features found, say, in chimpanzee tool behaviour.[9] These include (1) optimal adjustment to the task at hand, (2) the ability to behave adaptively with regard to a spatially and temporally removed environmental feature, (3) the ability to adjust behaviour to subtle variations in the environment, (4) the use of counter-strategies to neutralize prey defences, (5) an important role of play in the ontogeny of the behaviour, and (6) an important role for observation learning in the diffusion of the behaviour.

Of these features, the second is perhaps the most important from the viewpoint of the present work. In some of the cases reported by Beck, there was 'increased difficulty in solving the problem when the tool and the food were not in the same visual field'.[10] In other cases, however, chimpanzees carried the tools 'directly to mounds that were up to 90 m away and probably invisible from the point of tool selection'.[11] (Even more amazingly, some also carried spares – a behaviour which would seem to require a fairly complex mental operation of envisaging the possibility of the tool breaking down.) An instance of temporal remoteness is provided by baiting, as when herons place bits of bread on the water and seize the fish that become attracted to them.[12] Presumably the heron might have eaten the bread itself, but preferred to invest it for the sake of later consumption. As argued in Ch. 3 above, such behaviour – when observed in novel situations – is evidence of mental life and intentionality. Beck has an extensive and somewhat inconclusive discussion of the relation between tool behaviour and problem-solving by insight, as distinct from trial-and-error procedures. Although he strongly emphasizes the latter form of learning, he admits that there seem to be some cases which are 'resistant to explanation in terms of previous associative learning and experience'.[13] I believe, however, that the issue of insight should be dissociated from that of intentionality. Insight – the sudden perception of the possibility of doing things differently – requires intentionality, but the converse does not hold. Even behaviour that was originally acquired by imitation or by trial and error, may require intentionality in its performance.

To explain tool behaviour, and more generally behavioural innovation among animals, one may follow several strategies. First, one may take as explanandum a specific type of behaviour and explain it as the direct result of natural selection. One might imagine, say, that a point mutation or a series of such mutations led to the use by crabs of stinging anemones as tools for defensive and offensive purposes.[14] In the large majority of cases discussed by Beck, this simple kind of explanation does not seem plausible.

Secondly, then, one may approach the same kind of explanandum by arguing that the behaviour in question emerges by learning or invention, i.e. by trial and error or by insight. (I am focusing now on the emergence rather than the diffusion of the behaviour, so that cultural transmission and observation learning are not relevant.) It is then natural to look into the background conditions favouring or fettering the emergence of novelty. From Beck and Fagen it would appear that the following variables are important. (i) There must be time for object play and object exploration. This implies, among other things, that 'food shortages which force animals to switch to less preferred foods that are difficult and time-consuming to gather, process, or eat will cause decreased object manipulation. A decrease in object manipulation would decrease the probability of discovery of a novel tool pattern.'[15] Against this we should set the view that 'necessity is the mother of invention' often used to explain human tool behaviour. I return to this in section (c) below. (ii) The environment must be sufficiently varied to permit object exploration. Fagen suggests that the optimal temporal profile may involve 'an enriched rearing environment to develop versatility, then a period of deprivation and subsequent enrichment'.[16] This also implies that territoriality would tend to fetter the development of tool behaviour.[17] (iii) 'The potential innovator must be isolated from conspecific play companions',[18] for two reasons. First, if play companions are available, social play rather than object play may result. Second, companions may act as disincentives to innovation by stealing the inventions before the innovator can reap the fruits. This, however, really belongs to the next set of explanations, to which I now turn.

For, thirdly, one may take as explanandum the emergence of genes for playfulness and inventiveness rather than specific forms of behaviour. Fagen here makes the fundamental observation that 'questions of origin and of dissemination are independent at the level of mechanism, but not at that of evolution, since "theft" of useful discoveries from the inventor by genetically unrelated conspecific observers would tend to counterselect genetically based individual tendencies toward innovation'.[19] Since there are costs associated with playful behaviour, viz. the opportunity cost of food and mating possibilities foregone, the behaviour will not emerge unless it also has associated with it some fitness benefit. This benefit, however, will tend to be undermined if unrelated conspecifics immediately adopt the behaviour. Fagen briefly speculates about some ways in which this free-rider problem could be overcome: by innovators inventing so fast that imitators are always being left behind; by formation of play groups involving only individuals with two (recessive) alleles for invention; by secrecy, misinformation and concealment with respect to beneficial discoveries; by free exchange of discoveries among close kin coupled with restricted non-kin interactions; and by sharing discoveries with non-innovators in return for protection.[20]

Clearly, these mechanisms for facilitating invention differ in their evolutionary status. In particular, the mechanisms involving play groups and (to a somewhat lesser extent) kin groups are better at explaining the stability of genes for inventiveness than at explaining their emergence, since both of them presuppose interaction between individuals sharing those genes.[21]

These biological theories of tool behaviour and innovation suggest several questions that could also be relevant for the kind of technical change that forms the main topic of the present work. First, what is the proper explanandum of theories of technical change? Given that a firm at time t operates one technique or practice, should one try to explain what it will do at time $t+1$? Or should one rather set out to explain the amount of money the firm will invest in innovation? Secondly, are innovations favoured by abundance and prosperity, or by crisis and hardship? Thirdly, could some of the mechanisms by which evolution overcomes the free-rider problem also be relevant for economic theory? In particular, could the idea that innovators are protected by their own momentum – rather than by patents and similar arrangements – find application in cases where diffusion is less than instantaneous? This idea turns out to be important in many of the Nelson-Winter models. I cannot stress too much that biological analogies can only suggest questions, never provide answers. For the reasons set out in Ch. 2 and Ch. 3 above, evolutionary and intentional adaptation differ in crucial respects that invalidate many of the currently fashionable uses of biological models in the social sciences.

b. Eilert Sundt

In the first section of this chapter I dealt with models explaining technical change by *reduction* to natural selection: either the one-step reduction that explains specific tool behaviour as the result of natural selection, or the two-step reduction that explains genes for inventiveness by this mechanism.[22] I now turn to a series of models that try to understand technical change by some kind of *analogy* with natural selection. The analogy in all cases will be imperfect only: in fact increasingly so for the three models to be considered, linked to the names of Eilert Sundt, Richard Nelson and Sidney Winter, and Paul David.

Eilert Sundt (1817–75) was a theologian who devoted his life to pioneering work in demography and sociology.[23] He united a strong moralizing bent with an extremely forceful analytical and inquiring mind, a not uncommon combination at the time. His curiosity and analytical rigour often made him arrive at conclusions which he tended to resist on moral grounds, as when he found that certain strange courtship rituals in Western Norway could be explained by what we today would call the theory of decision under risk. After a visit to England in 1862 he came to be extremely impressed by Darwin's theory, and applied it to the ethnological study of artefacts such as houses and boats. Before his visit to England he had already published a long study of rural building customs, but later came to believe that these should rather be understood in an evolutionary framework. When in 1862 he gave a lecture on the type of fishing boats used in Northern Norway, he also, by way of introduction, offered the following remarks on the evolution of building customs:

> Even when people who set up new buildings did not intend to deviate from custom in any way, it could easily happen that some small variation arose. This would then be *accidental*. What was *not accidental*, however, is that inhabitants of the house and the neighbours should *perceive* the variations and form an opinion as to their advantages and inconveniences. And it is then not at all surprising that when someone later wanted to set up a new house, he would carefully *choose* that house to *imitate* which seemed to him the most useful. And when the idea of improvement in a definite direction had first emerged, someone more clever could then take a further step and actually envisage and carry out another improvement.[24]

Later in his talk Sundt elaborates on these ideas as they impinge on his chosen topic, the technology of fishing boats:

> A boat constructor may be very skilled, and yet he will never get two boats exactly alike, even if he exerts himself to this end. The variations arising in this way may be called *accidental*. But even a very small variation usually is noticeable during the navigation, and it is then *not accidental* that the seamen come to *notice* that boat which has become improved or more convenient for their purpose, and that they should recommend this to be *chosen* as the one to *imitate* .... One may believe that each of these boats is perfect in its way, since it has reached

perfection by one-sided development in one particular direction. Each kind of improvement has progressed to the point where further developments would entail defects that would more than offset the advantage . . . . And I conceive of the process in the following way: when the idea of new and improved forms had first been aroused, then *a long series of prudent experiments,* each involving extremely small changes, could lead to the happy result that from the boat constructor's shed there emerged a boat whose like all would desire.[25]

In this quasi-Darwinian model of technical change, Sundt has incorporated the following ideas. (i) The variations are *random*, at least initially. (Later the variations are produced deliberately, by experimenting with small changes, although it is not clear whether Sundt means that the direction as well as the rate of variations is subject to choice.) (ii) The variations are typically very *small*. (iii) The result of variation-cum-selection is a *local* maximum. (iv) The imperfection of the constructor – his inability to make perfect copies – is a condition for the ultimate perfection of the end result. (v) The choice of boats is a fortunate one, in that here the sensibility of the end result to small variations is large. Sundt is mistaken, however, when he states that this is a model fully analogous to natural selection. His emphasis on perception and subsequent choice shows that his is a model of artificial rather than natural selection, warranting filter-explanations rather than functional explanations of technical change. As observed in Ch. 2, artificial selection differs from natural selection in that it permits waiting and the use of indirect strategies, unlike the myopic opportunism of biological evolution. Animal breeders, for instance, practice waiting when they keep certain options (genes) open for later use, should the need for them arise, instead of selecting exclusively for the current maximization of whatever they are trying to maximize. Similarly the deliberate choice of inferior alternatives for the purpose of attaining a global maximum later on can be realized by artificial selection, although in the particular case discussed by Sundt the fishermen would hardly lend themselves to this procedure.

Still, artificial selection also differs from intentional adaptation in that it is restricted to choice between randomly arising variations. Unlike natural selection, artificial selection can take account of the long-term consequences of current choice, but unlike intentional adaptation it is constrained to accept or reject the small variations that are thrown up by the random mechanism. True, the rate and possibly the direction of these inputs to the selection process may be shaped by intentional action, as suggested by Sundt towards the end of both the quoted passages, but as long as they are constrained to be small there is no presumption that any alternative, however distant, can be reached through a series of stepwise changes, even when some of the latter are allowed to be unfavourable. It is not clear why these induced variations are constrained to be small in the case of boat improvement, but clearly this holds true of animal and plant breeding.

The foregoing enables us to detect an ambiguity in the phrase 'local maximization' that has been repeatedly invoked here.[26] First, we have a local maximum if there is some neighbourhood around the state in question in which there is no alternative that is better according to some criterion of optimality – which might be short-term or long-term. Secondly, we have a local maximum if no further improvement can take place without first passing through a temporary deterioration. The first notion does not involve time, the second essentially does. Myopia, similarly, may either imply the inability to perceive alternatives distant in the possibility space, or the inability to perceive (or be motivated by) long-term consequences of current choices within that space. Natural selection is myopic in both senses, intentional choice in neither, and artificial selection as conceived of here is myopic in the first but not in the second. (The fourth and remaining category is instantiated by a free agent subject to weakness of will.) The relative importance of the one or the other form of myopia in restricting action will depend on the structure of the possibility space and on the nature of the causal chains involved. Yet it should be emphasized that the 'local maximum trap' characteristically associated with natural selection is due to the inability to act on long-term consequences, not to the smallness of the changes. If, *per impossibile*, natural selection were able to accept unfavourable mutations and reject favourable ones, the global maximum might still be unattainable, but not because evolution had become trapped in a local maximum.

c. Nelson and Winter

I have singled out Joseph Schumpeter as probably the most influential single writer on technical change. Yet later economists have had difficulties in putting his theory to explanatory use. 'The trouble with Schumpeter's theory is that it is descriptive rather than analytical', says one writer.[27] Two others state that 'Many economists agree with the broad outlines of Schumpeter's vision of capitalist development, which is a far cry from growth models made nowadays in either Cambridge, Mass., or Cambridge, England. But visions of that kind have yet to be transformed into theory that can be applied to everyday analytical and empirical work.'[28] The latter statement can no longer be regarded as true. In a long series of important articles Sidney Winter and Richard Nelson have worked out in considerable analytical detail, with an eye to empirical application, a rigourous version of Schumpeter's theory. True, their models derive as much from Herbert Simon as from Schumpeter,[29] and we shall see that their repeated insistence[30] on the Schumpeterian nature of their theory is not fully justified. They are closer to Schumpeter when they reject the neoclassical notions of *maximizing rationality* and *equilibrium* than when they offer *search* and *selection* as their own alternatives. This and other differences notwithstanding, their work may be seen as a 'creative development' of Schumpeter's, with more stress perhaps on the first part of that phrase.

In addition to the joint work by Nelson and Winter, the latter has written a series of articles about the methodological foundations of their evolutionary alternative to neoclassical economics.[31] One of his contributions is of particular interest and importance: the demonstration that the neoclassical notion of maximizing involves an infinite regress and should be replaced by that of satisficing. The argument appears to me unassailable, yet it is not universally accepted among economists, no doubt because it does not lead to uniquely defined behavioural postulates. Satisficing involves, as explained earlier, an arbitrary restriction of attention: a restriction of the set of possibilities to be scanned that cannot be justified by rational argument. Winter argues that satisficing is inescapably present in all purposive behaviour, rather than being a mode of choice we can choose to adopt as an alternative to maximizing. His classic statement of this view is worth quoting at some length:

> It does not pay, in terms of viability or of realized profits, to pay a price for information on unchanging aspects of the environment. It does not pay to review constantly decisions which'require no review. These precepts do not imply merely that information costs must be consid-

ered in the definition of profits. For without observing the environment, or reviewing the decision, there is no way of knowing whether the environment is changing or the decision requires review. It might be argued that a determined profit maximizer would adopt the organization form which calls for observing those things that it is profitable to observe at the times when it is profitable to observe them: the simple reply is that this choice of a profit maximizing information structure itself requires information, and it is not apparent how the aspiring profit maximizer acquires this information, or what guarantees that he does not pay an excessive price for it . . . .

The satisficing principle is not, of course, a theory of the firm . . . . To the extent that selection considerations lend it support, it is largely because the explicit recognition that conceivable modes of behaviour remain unconsidered provides an escape from the *reductio ad absurdum* of information costs. At some level of analysis, all goal seeking behavior is satisficing behavior. There must be limits to the range of possibilities explored, and those limits must be arbitrary in the sense that the decision maker cannot *know* that they are optimal.[32]

As observed by Winter, the idea of satisficing is compatible with a wide range of behavioural patterns, and does not in itself provide a theory of entrepreneurial behaviour. The theory only tells us that at *some* level of analysis, the decision must invoke satisficing, not *which* level this is (or should be). We shall see below how, in their various models, Nelson and Winter invoke specific behavioural postulates of satisficing.

Nelson and Winter have developed their arguments in a series of related, though subtly different, models. They address among others the following problems: to explain the rate of technical change, to explain market structure as an endogenous variable, to explain the factor bias of technical change, to explain the relative importance of innovation and imitation in technical change. These are problems also discussed in neoclassical models, but the conceptual tools used by Nelson and Winter are wholly different from those of neoclassical economics. First, they completely reject the notion of the production function as the proper conceptualization of the state of technical knowledge. It follows fairly easily from the general theory of satisficing that firms cannot be expected to have at their finger tips a detailed knowledge of techniques other than those which they are currently using. Rather, 'surveying available knowledge is like surveying a landscape on a hazy day. Some things are close enough to be seen clearly, others remote enough to be totally invisible. But the intermediate conditions cover a broad range and are complex to describe.'[33] To develop these intermediate techniques will require time, effort and expense – although it is impossible to say in advance how much. The principle of satisficing, moreover, dictates 'that investment in the quest for understanding be deferred until there is a symptom of trouble to deal with.'[34] This implies that if for some reason firms are driven to change technique, they will have to search for another one – with no guarantee that they will find one that is better than the one which is currently used. In fact, the search is best modelled probabilistically, by assuming that the probability of finding a superior technique – by innovation or imitation – is a function of the amount invested in the search.

The element of *search* is one basic notion in the Nelson-Winter models. The other fundamental idea is that of *selection* of firms by the market. Firms which happen to find better techniques or which happen to use better rules for searching than others, will expand relatively more. (Here 'better' must be taken strictly in an *ex post* sense: *ex ante* all behaviours are assumed to be equally good or 'satisfactory'.) Exactly how the expansion comes about depends on various features of the market: on the number of firms, on the degree of aggressiveness, on the availability of external financing, etc. To get a firmer hold on these ideas, it will be useful to look in more detail at some of the families of models employed by Nelson and Winter. First I shall briefly discuss a family of equilibrium models, and then the more canonical non-equilibrium models. Within the latter there is a distinction to be made between an early attempt to explain technical change by postulating satisficing at the level of current technique and the more recent models in which firms satisfice on their R & D policies.

All models assume that there is a population of firms, and that at each moment of time each firm can be characterized by a number of state variables, such as amount of capital, current technique (or current productivity), research policy, etc. Moreover, the models define certain behavioural postulates concerning the research and investment activities of the firms. Finally, there is a set of conditions relating factor prices and product prices to factor demand and product supply. Not all models include all of these variables and conditions, but each model has some state variables, some behavioural postulates, and some market conditions of this kind. Given the values of the state variables at time t, the other postulates and conditions enable us to define a probability distribution over the set or sets of state variables that can obtain at time $t+1$. The model, in other words, can be seen as a Markov process with constant transition probabilities. The reason why the model is probabilistic rather than deterministic is that the result of the search processes is stochastic. The models have uncertain and unpredictable technical change at their very core.

These models can be studied in two very different ways. First, one may try to obtain explicit analytical results concerning the evolution of the firms over time. Winter in one early analysis modelled technical change in a way closely related to what in Ch. 2 above was referred to as an absorbing Markov chain.[35] He showed that, given a number of quite restrictive assumptions, an equilibrium state would eventually occur with probability one and be indefinitely maintained thereafter. The transition rules of the model were particularly simple: profitable firms expand without changing technique, unprofitable firms contract while changing technique, and firms making zero profits behave unchangingly. (In addition there are rules for entry into the industry.) The process generated by these transition rules converges to a state in which all firms make zero profits. Although the result is theoretically interesting, the model is too devoid of economic content and structure to tell us much about the behaviour of real economies. A later Nelson-Winter model discussing the factor bias of technical change in a similar equilibrium framework goes somewhat beyond this first attempt, yet is explicitly proposed as a methodological exercise only.[36]

Secondly, however, one may build models too complex to be studied by analytical methods, and rely instead on computer simulations. This is the procedure of the – to my mind – more interesting Nelson-Winter models, to which I shall now turn. I postpone the discussion of the status of the conclusions obtained by simulation studies until I have offered a sample of the models and their results.

An early and fairly simple model was proposed by Nelson, Winter and Schuette.[37] It embodies the following notions. (1) *Satisficing* on techniques. Firms retain their current technique if their gross return on capital exceeds a certain level. (2) *Induced search*. If the gross return falls below this level, firms engage in local search for new techniques or in imitation of other firms. (3) *Profitability testing.* A firm adopts a new technique turned up by the search if it promises to yield higher return on capital. (4) *Investment.* Firms expand by investing their gross returns, after subtraction of depreciation and required dividends. (5) *Entry.* Entry is free, but regulated by assumptions which entail that it is relatively infrequent. (6) *Labour market.* The wage rate is in part endogenous to the model, being influenced by the behaviour of the industry as a whole, and in part subject to an exogenous trend. As usual, the laws of motion are probabilistic and depend on the probability of a successful search.

For the computer simulations four parameters were varied systematically, each taking on two values so that a total of 16 cases were considered. These were: the ease of innovation through local search, the strength of imitation relative to innovation, the size of required dividends, and the direction of the local search (neutral versus laboursaving). In each computer run all firms are assumed to have identical values of all these parameters, so that their importance only shows up in comparison between runs. Each case was run for 50 periods in the computer, corresponding to Robert Solow's study of the US economy from 1909 onwards.[38] Solow argued that the data could be explained neoclassically, by assuming that the economy is working with an aggregate production function subject to neutral technical change. Against this the authors argue (i) that the historical data fit almost equally well with some of their computer runs, and (ii) that if one tries to give a neoclassical explanation of the values generated by simulation runs, one succeeds remarkably well – in spite of the fact that these values are generated by a process that does not embody the neoclassical notions of maximizing or the production function. 'The fact that there is no production function in the simulated economy is clearly no barrier to a high degree of success in using such a function to describe the aggregate series it generates.'[39] And they conclude that 'this particular contest between rival explanatory schemes should be regarded as essentially a tie, and other evidence consulted in an effort to decide the issue'.[40] Such evidence, the authors imply, must be found at the micro-level, i.e. in the theory of decision-making by firms.

Comparison between the runs shows up the importance of the parameters, occasionally in surprising ways. A typical piece of reasoning is the following. It turned out that a high value of the first parameter, with innovation relatively easy and less constrained to be local, goes together with a high capital-labour ratio (in period 40 of the run). Why should this be so?

> On reflection, one possible answer to this question seems to be the following: The general direction of the path traced out in input-coefficient space does *not* depend on the localness of the search. However, the rate of movement along the path is slower if search is more local. Therefore, given that the path is tending toward higher capital-labor ratios (as a consequence of the level chosen for R^{41} and the neutrality or labor-saving bias of the search), the capital-labor ratio that results after a given number of periods is lower when search is more local . . . . Another possible answer is more 'Schumpeterian'. A high rate of technical progress may produce a high level of (disequilibrium) profits, which in turn are invested. The resulting increase in the demand for labor results in a higher wage, and deflects the results at profitability comparisons in the capital-intensive direction.[42]

A later family of models capture more complex interactions between firms.[43] Their most important feature is that firms are now assumed to satisfice on investment in R & D. In other words, the search for new techniques is not triggered by adversity, as in the model set out above, but is carried on as a matter of routine. Some firms do both research and imitation, while others only imitate. As usual, the results of search and imitation are stochastic only: a large investment in R & D may bear small fruits, and a small investment inordinately large fruits, depending on chance. In most of these models innovation is 'science-based', i.e. it draws on an exogenous increase in 'latent productivity'. In one model, however, the authors also consider a 'cumulative technology' case in which a firm innovates by making incremental improvements in its own current technique, not by drawing on new knowledge created outside the industry.[44] Among other conditions that vary in the computer simulations are the following. The number of firms in the population may be 2, 4, 8, 16 and 32 firms. There is also a parameter for degree of aggressiveness: a non-aggressive industry state obtains if firms require higher profits to invest as they grow larger, meaning that they take into account that they are large enough to depress the price by further expansion. Also allowed to vary are the ease of external financing, the ease of imitation, and the rate of growth of latent productivity. These are all inter-run differences, whereas, to repeat, the distinction between innovators and imitators is an intra-run difference in these models.

It is not possible to summarize here all the intriguing results obtained by the authors in working with these models. I shall limit myself, therefore, to some conclusions that appear especially relevant for the present purposes. When summarizing these results, I shall not at every step indicate the precise features of the models in which they are derived: this the interested reader will have to look up for himself.

A constant theme in the models is the 'Schumpeterian hypothesis' of a trade-off between competitive behaviour and progressiveness. In their discussion of this issue the authors make several valuable distinctions. First, one may define competition either structurally, by the number of firms in the industry,[45] or behaviourally, by the degree of aggressiveness. Secondly, we may distinguish between three senses of progressiveness, according as the standard chosen is the level of average productivity, the productivity of best practice techniques, or the price of the product. In the case of science-based technical change, the number of firms but not the degree of aggression made a difference to progressiveness. More specifically, a four-firm industry compared to a sixteen-firm industry had a higher average productivity, but also higher prices, since the lower costs were more than offset by the higher markup over cost.[46] In the cumulative technology case, however, aggressive behaviour had a negative effect on progressiveness in all three senses, since the higher cost failed to be offset by lower markups.[47]

Another constant theme is the relative fate of imitators and innovators. In one set of models the innovators do better when competition is restrained, and the imitators better when it is aggressive. Once again a quote may be useful to give a flavour of the reasoning invoked to explain the findings:

> The explanation of this phenomenon . . . resides in the following. In this model an imitator can never achieve a higher productivity level than the best of the innovators. If he matches an innovator's productivity he will have higher profits, because he does not incur the innovative R & D costs. But if he stops his output growth at reasonable size, pressures on the innovating firm to contract are relaxed, the R & D budget is not eroded, and there is a chance of recovery for the

innovating firm . . . But if the large imitating firm continues to grow, it forces the innovating firm to continue to contract. As the innovator's R & D budget contracts, the chances of an innovative success that will spark recovery diminish, and the expected lead time before the big imitator imitates diminishes as well.[48]

In a different set of models, there arises a puzzle because researchers do as well as imitators in all but the 32-firm cases, and yet there seems to be decisive evidence that research does not pay.[49] The explanation is that research turns out to be a 'slack' activity, something the researcher can afford by virtue of being a rich oligopolist, but that really does not pay. With a larger number of firms, investment must be financed externally to a greater extent, and hence depends on the net worth of the firm, which is negatively affected by R & D outlays. This, however, does not necessarily lead to the elimination of the innovators. Rather, 'a different type of "equilibrium" sharing of the market arises – one in which the imitators are large enough to desire markups high enough so that the researchers can break even'.[50]

Against the background of this random sample of results generated by the Nelson-Winter models, I want to discuss three issues. First, in what sense can they be said to be 'Schumpeterian'? Secondly, what is their relation to the biological theory of evolution? Thirdly, what is the explanatory status of the simulation methods?

Schumpeter is present in the models in several ways. For one thing, the authors discuss and, broadly speaking, embrace the Schumpeterian hypothesis of a trade-off between the static efficiency usually associated with competition and the dynamic efficiency of restrained competition. Also, and more importantly, they insist on the Schumpeterian idea that competition is a process rather than a state. In neoclassical models, competition is little more than price-taking behaviour, whereas in the everyday sense of the term it is a process with identifiable winners and losers. Moreover, in economic life as well as in sport sheer luck plays an important role in distinguishing winners from near-winners, 'although vast differentials of skill and competence may separate contenders from noncontenders'.[51] There is no doubt that this is a genuinely Schumpeterian feature of the models. On the other hand, the crucial notion of satisficing derives from Herbert Simon rather than from Schumpeter. The importance of large discontinuous innovations in the Schumpeterian scheme is not found in the Nelson-Winter models, nor is the emphasis on supernormal ability and energy as a condition for profit-making. I also believe that the idea of firms steadily tracking the growth of latent productivity is somewhat foreign to Schumpeter's view of technical change. In this respect the Frisch-Goodwin models are more faithful, since at least they incorporate the element of discontinuity. What was unacceptable in the latter models, viz. the automatic and deterministic transformation of latent productivity increase into innovation, is of course not a feature of the Nelson-Winter models at the microlevel. Yet at the aggregate level the result may be, in some versions, that the population of firms tracks the growth of latent productivity very closely – and continuously.

The very phrase 'evolutionary theory of technical change' suggests of course a biological analogy. Nelson and (especially) Winter often discuss the relation of their theory of technical change to the theory of biological evolution.[52] They are very sensitive to the many disanalogies between the two theories, and at no point do they fall into the trap of illegitimately transferring the conclusions from one to the other. The broad analogy resides in the use, in both theories, of stochastic variation and subsequent deterministic selection. Among the disanalogies, the most important is discussed separately below, viz. the lack of a notion of equilibrium in the Nelson-Winter models. If we look to the finer grain, the following differences also appear. First, although the variations are typically fairly small, they need not be totally random in the models of technical change. For instance, over and above the fact that factor prices may influence the selection of new technologies, 'it is plausible that a rise in the wage rate should induce a shift in the distribution of the firms' search efforts'.[53] Moreover, the stream of variations or 'mutations' need not be constant, as it is in the theory of biological evolution. In the Nelson-Winter-Schuette model, the rate of mutations is zero when no mutations are needed, and only becomes positive when gross returns fall below some critical level. We are dealing with a *satisficing machine* differing both from the locally maximizing and the globally maximizing machines discussed in Ch. 2 above. (In biology, the satisficing machine would not *stricto sensu* violate the central dogma of no feedback from environment to genome, but it would clearly violate the spirit of that dogma.) On the other hand, in the later Nelson-Winter models firms satisfice on R & D, inducing a constant rate of 'mutations'.

148 THEORIES OF TECHNICAL CHANGE

Secondly, the selection process in economics differs from that of biology. Firms, like organisms, can die, although in most of the Nelson-Winter models they only dwindle towards extinction without ever totally disappearing. They cannot, however, reproduce themselves. When their features – the techniques or search rules that make them successful – spread in the population of firms, it is partly by expansion and partly by imitation. Imitation is more like artificial than natural selection, in that it involves a subjective element of perception and not only the objective element of differential survival. Expansion, by contrast, is somewhat more analogous to the workings of natural selection. Yet this mechanism means that the success of a technique or of a search rule cannot be measured by the number of firms adopting it – rather the firms must be weighed by their size.

The explanatory status of the Nelson-Winter models is puzzling and intriguing. They exhibit largely *non-equilibrium* and even *non-steady-state* behaviour, which means that it is difficult to establish explanatory connections between the variables. Let me begin with the disequilibrium aspect of the models, which is strongly emphasized by the authors. The most general statement seems to be the following:

> This kind of model can have an equilibrium with neoclassical properties, but it is also possible to explore the disequilibrium properties of the model, and indeed to set the context such that equilibrium does not obtain over the entire relevant time span. While firms find better techniques, there always may be still better ones to be found. Profitable firms expand, unprofitable ones contract. But the system need not drive out all but the most efficient techniques or decision rules. Changes in the 'best' techniques known by firms and in the external environment of product demand and factor supply conditions may be *sufficiently rapid relative to the speed of adjustment of the overall system* that a wide range of behavior can survive at any time.[54]

In the phrase that I have italicized the authors invoke an argument similar to the one used in Ch. 2 above to refute Arthur Stinchcombe's defence of functional explanation. The argument shows that one cannot in general – without specific evidence – assume that observed behaviour can be explained by its properties of rationality and optimality. Unlike the case of natural selection in biology, the economic environment may change too rapidly for inefficient firms to be eliminated, since what is efficient at any given time depends largely on the environment. Observe the two-pronged nature of the Nelson-Winter argument. On the one hand, they invoke general considerations of satisficing to argue that the firm does not and cannot maximize *(ex ante)* in any subjective sense. On the other hand, they point to the rapidly changing environment to argue that selection may not have the time to bring about *(ex post)* optimality in the more objective sense. The coexistence of efficient and inefficient firms at all points of time is a decisive reason for distrusting the biological analogy.

Yet even if the models never are in equilibrium, they may have steady-state properties. For instance, the relative proportion of efficient and inefficient firms may after some time settle down to a fairly stable value. In the Nelson-Winter-Schuette model the authors remark that 'the reason for focusing on values observed late in the run is to allow plenty of time for the different parameter settings to display their distinctive influence on the industry state'.[55] Firms, in this perspective, may rise and disappear for stochastic reasons, and yet the patterns may remain quite stable over time. In such cases we can invoke the parameters to *explain* the industry state, in accordance with what I referred to in Ch. 1 above as 'simultaneous causation'. Equilibrium models permit explanations of phenomena in terms of their actual – possibly also intended – consequences. Steady-state models permit causal explanations of phenomena in terms of the exogenous variables that influence the patterns in which the system ultimately comes to rest. As also explained in Ch. 1, the derivation of the steady state may be quite complex and require the analysis of a large number of opposing mechanisms, the net effect of which may be difficult to predict *ex ante*. The Nelson-Winter approach, in fact, is very close to the Tocquevillian methodology outlined earlier.

Finally, in some of the models there is not even a steady-state behaviour of the population of firms. For instance, in the models that initially have 16 or 32 equal-sized firms, there is a strong upward trend in concentration that, at the end of a 25-year simulation, has not arrived at a steady state. The authors conclude that this upward trend is a firm result of the simulation.[56] This may well be justified by their intimate knowledge of how the models work, but for an outsider without this knowledge the possibility of cycles cannot be excluded. In other words, to the upward trend towards concentration there might conceivably succeed a downward trend towards a more atomistic industry structure. To invoke Tocqueville once more, the transient effect of a given parameter change may differ – not only in size, but even in direction – from the steady-state effect.

This concludes my discussion of the Nelson-Winter models. It will be obvious from the general tenor of my discussion that I am very sympathetic to their approach, and equally obvious that I can claim no competence to evaluate its soundness for the purpose of empirical research. The image of the economy as a constant flux with certain moderately stable patterns clearly has the advantage of realism over the neoclassical vision. It remains to be seen whether the evolutionary school will be able to cash in on this advantage for explanatory ends.

d. Paul David

The last evolutionary theorist to be discussed here is Paul David. Among the writers that I have singled out in this chapter, he is also the least evolutionary, in the sense that his work is less closely related to the biological theory of evolution. Yet he himself refers to his work as an evolutionary alternative to neoclassical theory, and it will become clear that this is indeed an appropriate description. David's work on technical change is presented in his *Technical Choice, Innovation and Economic Growth,* in which he has collected a number of studies in technological history and added a long initial chapter on 'Labor scarcity and the problem of technological practice and progress in nineteenth-century America'. The following exposition draws exclusively upon this chapter. Although explicitly historical in its subject matter, the chapter also sketches a general theoretical framework for the study of technical change. The basic fact to be explained is the link between labour scarcity and mechanization in nineteenth-century America as compared to the English development during the same period. David critically discusses and rejects the neoclassical explanations of this factor bias proposed by Fellner and Kennedy (Ch. 4 above), and then elaborates an alternative, 'evolutionary', explanation. Surprisingly, he also rejects the work of Nelson and Winter as being 'fundamentally neoclassical in spirit'.[57]

In order to understand the logic of David's own alternative, we may first observe that he is heavily influenced by Kenneth Arrow's seminal article on 'Learning by doing'. That article in turn had as one point of

EVOLUTIONARY THEORIES   151

departure the so-called 'Horndal effect'. 'The Horndal iron works in Sweden had no new investment (and therefore presumably no significant change in its methods of production) for a period of 15 years, yet productivity (output per manhour) rose on the average close to 2 % per annum. We find . . . steadily increasing performance which can only be imputed to learning from experience.'[58] This kind of productivity increase is *local* and *neutral:* local in that it involves incremental and adaptive change only, and neutral in that it saves equal proportions of all factors of production. David's basic proposal is to explain the overall rate and direction of technical change as the sum-total of many local-cum-neutral changes.

Before we survey the details of David's analysis, I shall briefly review what was said or implied in the preceding section about the production function. It follows from the Nelson-Winter approach that firms have no direct access to any other methods than the one which they are currently using. If they are induced to search for new techniques, they have no certainty of finding superior practices, only a certain probability of doing so. In some of their models the search also involves costs, so that investment in R & D implies that firms buy a probability of sampling from a distribution which includes some superior techniques. There is no room here for the notion of a production function involving a large number of practices to which firms have immediate and free access. David (and also Nathan Rosenberg in closely related work)[59] criticizes the notion of the production function from a somewhat different angle. He assumes that firms at any given time have immediate and free access to a small number of practices and any linear combination thereof. In Fig. 3 below (taken from David) there are two such practices, with capital-intensities corresponding to the slopes of the rays α and γ from the origin. The 'available process frontier' $APF°$ consists of all linear combinations of the two basic practices. The 'fundamental production function' $FPF°$ consists of all the practices that might be made available by drawing on existing scientific and technical knowledge. They are not immediately and costlessly available, since research and development are needed to bring them out of their latent state. The economic relevance of these concepts is explained as follows:

[Suppose] that prior to an increase in the wage-rental price ratio (shown as the change in the price line from pp to p'p'), producers were

Fig. 3. Technical innovation viewed as substitution.

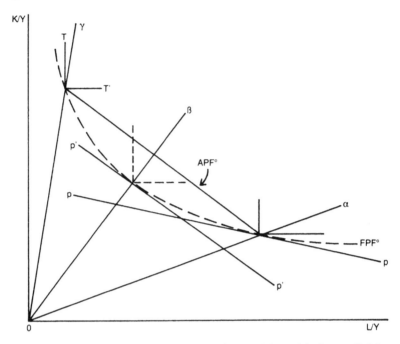

content to minimize production costs by working with the available α-technique. An immediate consequence of the indicated rise in the relative price of labor would be to render them indifferent, on the same considerations, between staying with α or switching to γ. And a further marginal change of relative wage rates in the same direction would decide the question in favor of the more capital-intensive direction. But an alternative response could also be elicited from some producers by this factor-price change: the allocation of resources to the actual development of a β-technique.[60]

This shows that the line between innovation and substitution cannot be drawn with the precision suggested by the neoclassical apparatus. To move from the α-practice to the γ-practice is substitution without innovation, whereas the switch from the α-practice to the β-practice is a form of substitution that also involves innovation. And lastly there is innovation proper: innovation without substitution. Consider Fig. 4 (from David). Here the price switch from pp to p'p' has induced a shift from the α-practice to the β-practice. David then assumes that there occurs innovation in the form of local-neutral technical progress, so that the best practice moves along the β-ray towards the origin. (In fact, as explained below, it does not move *along* the ray, but in a random way close to it.) As the β-practice moves towards the origin, the available process frontier tilts in the labour-saving direction, becoming ultimately APF'. If at this point the price ratio changes from p'p' back to pp, there will be no reversal to the α-practice, since the improved β-practice has become superior at this ratio as well.

We shall see in a moment how this model may be put to do explanatory work. Let us first pause, however, to consider the implications for the neoclassical production function. We have seen that David's approach underlying Fig. 3 implies a mild critique of the production function, by suggesting that substitution may involve an element of innovation. The apparatus presented in Fig. 4 contains the elements of a more radical critique. It suggests, metaphorically speaking, that the production function isn't there when one is not looking at it. Or in more precise language, the production function can hardly be a useful concept unless it has more than a merely transient existence. This implies at least the following requirements. (a) If the function has non-constant returns to scale, and one moves from one isoquant to another, then the first should still be there if for some reason one wants to go back. (b) If a firm currently occupies one position on the unit isoquant (assuming from now on constant returns to scale) and then moves to another position, then the first should still be there if for some reason it wants to move back. (c) Minimally, if the firm operates at some point on the unit isoquant, then simply 'being there' should not have the effect of changing the practice.

As briefly observed in Ch. 1 above, David's critique of Robert Fogel amounts to a denial of the first requirement, since there typically are irreversible economies of scale that permit more efficient production if again one returns to the smaller scale. The non-fulfilment of the second requirement is often expressed by some such phrase as 'movements *along* the production function cannot be distinguished from movements *of* the production function'.[61] And the model shown in Fig. 4 shows that even the third requirement is violated in a model with 'learning by doing'. Operating within a set of possibilities has the effect of changing the set itself. It is clear that if these objections are accepted, the notion of the production function may lose its analytical cutting edge. How much it will

154   THEORIES OF TECHNICAL CHANGE

lose depends on the relative speed of the processes involved and on the purpose of the analysis, since it may be possible to treat a slowly moving frontier as being at rest for the purposes of short-term adjustment. Another possibility – suggested by the analogous problems that arise in the case of the utility function[62] – would be to index the production function by the currently operated practice, and to assume (a) that firms choose the practice for time $t+1$ according to the function corresponding to the practice operated at time t, and (b) that a function corresponding to the $t+1$-practice also arises in a well-determined way.

Fig. 4. Localized learning and the global bias of technological progress.

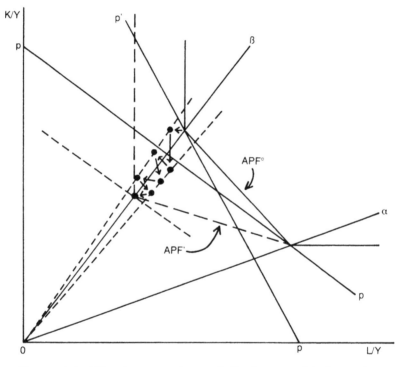

The model of Fig. 4 can now be used, firstly, to explain how a factor-bias in technical change can arise without any changing trend in factor-price ratios. Following Habakkuk[63] and David, we assume that in nineteenth-century technology movements along the capital-intensive β-ray were inherently faster than movements along the α-ray. This is tantamount to saying that the ease of innovation is inherently greater in one end of the spectrum than in another. We do not, however, have to require entrepreneurial knowledge about the relative ease of innovation, since firms innovate locally-neutrally in whatever part of the spectrum they happen to find themselves. (Thus David's model does not have to confront the problem of microfoundations that we observed in connection with the Kennedy model discussed in Ch. 4.) We assume, moreover, that the factor-price ratio shows no overall trend, but fluctuates around a constant level. It then follows that technical change will on the whole be labour-saving, since more innovation takes place when prices favour practices in the capital-intensive part of the spectrum than when they encourage labour-intensive methods. Equal time with high and low wage-profit ratios does not imply equal amounts of capital-intensive and labour-intensive innovations, since we assume that more innovation occurs when the economy is in a high-wage phase.

The most important application of the model, and the purpose for which it was constructed, is the explanation of the different biases of technical change in America (a high wage region) and Britain (a low wage region) during the nineteenth century. If in the former technical change progresses along the β-ray and in the latter along the α-ray, then their shared fundamental production function (assuming each country to have free access to the basic technical knowledge of the other, through migration of engineers etc.) will shift towards the origin in a labour-saving direction, since the movement along the β-ray is swifter than along the α-ray. We will then observe two things. Cross-temporally, we will see British producers engage in labour-saving innovations, since they have access to a 'production function' which shifts in the labour-saving direction. Cross-sectionally, we will observe that British practices at any given time are less capital-intensive than the American ones, since at the British factor prices the β-practice will rarely if ever be the optimal one. (The case here is different from the one considered above, since we are not – as in Fig. 4 – keeping the α-practice constant as the β-practice moves towards the origin.)

This is an explanation of how a high wage-profit ratio goes together with a labour-saving bias in technical change. The demonstration hinges on the assumptions of (a) rational substitution (or substitution-cum-innovation), (b) local-neutral technical progress, and (c) inherently faster local progress in the more capital-intensive part of the spectrum. On this basis David reaches the conclusion that neoclassical theorists in vain

156 THEORIES OF TECHNICAL CHANGE

tried to prove by the more direct method of showing the rationality of the bias itself. I have no comments on assumptions (a) and (c), but there is some methodological interest in David's argument for the crucial assumption (b).

In the first place David argues for the neutrality of technical change from the Principle of Insufficient Reason. Since there is no reason for believing that learning by doing has a consistent bias in any direction, we may model it as a random walk in which a shift to the left (in the labour-saving direction) is just as probable as an equally large proportional shift downwards (in the capital-saving direction). This implies that technical change can be expected to be neutral, i.e. that the mean will lie on the ray to the origin. But since the variance of an unrestricted random walk is uncomfortably great, and in fact increasing with the number of steps, David adds that the process of local-neutral change is constrained to operate within the 'elastic barriers' represented by the two stippled lines on either side of the β-ray in Fig. 4. The core of David's theory is probably the technological justification of this postulate. He argues, to put it very briefly, that there is 'technical interrelatedness (complementarities) among the sub-activities' involved in any given practice, so that a factor-saving change in one sub-activity calls forth a factor-using change in another.[64] I have no competence whatsoever to discuss the empirical adequacy of this idea, but clearly this is the place where the historian of technology would want to focus his attention.

The reader will recall that Nelson and Winter impute to the firm no certain knowledge about any practice but the one it currently operates. David stops short of imputing knowledge of the whole neoclassical production function, yet he does assume that the firm has access to several distinct practices and chooses rationally between them. It might seem, therefore, that David's departure from neoclassical theory is less radical than that of Nelson and Winter. I for one am prepared to argue this, while also adding that this contest in heterodoxy has a somewhat limited interest. It may be instructive, nevertheless, to look at David's argument for the view that the work of Nelson and Winter is 'fundamentally neoclassical in spirit'.[65]

David characterizes his own approach as evolutionary *and* historical. While agreeing that the Nelson-Winter theory also is evolutionary, he denies that it is historical. Because of this lack of a 'genuinely historical conception of economic growth as an irreversibly evolutionary development', their model is not an alternative to the neoclassical tradition, from which it differs only in the conception of micro-economic behaviour. In David's explanation of what he means by a 'historical model', the key passage is the following: 'unlike the large class of dynamic and static economic models inspired by classical mechanics, the future development of the system depends not only on the present state but also upon the way the present state itself was developed'. The Nelson-Winter approach does not satisfy this requirement, since they base their theory on Markov processes, which are 'probabilistic analogues of the processes of classical mechanics'. This refers to the fact, explained in Ch. 1 above, that in a Markovian model the transition probabilities depend only on the current state and not on earlier states of the system.

It follows directly from the discussion in Ch. 1 that – and why – this argument is wrong. David makes a case for hysteresis in a substantial sense, as if the past could have a causal influence on the present not mediated by a chain of locally causal links. In this he follows Georgescu-Roegen, whom he quotes with approval. But, as argued earlier, this view is contrary to all canons of scientific method. It may sometimes be convenient to model a process as if the past could have a direct impact on the present, but if one rejects action at a distance one must believe that this can only be a useful fiction. If a model is historical in David's sense, it is a distinct drawback, not an asset. Whenever one has a 'historical model', there should be a challenge to go further and eliminate the history by transforming it into state variables. And of course the fact that the local progress operates within an environment that is shaped by past progress is neither here nor there. It is just as true of the Nelson-Winter theory as of David's own. Nor would it be true to say that the Nelson-Winter models are insensitive to differences in the initial conditions. On the contrary, they show that even with identical initial conditions further developments may diverge, and *a fortiori* this must hold with different initial settings. I conclude that David's methodological objections to the Nelson-Winter theory are misguided, a conclusion that is only reinforced by his strange contrast between 'Darwinian' and 'Mendelian' theories of technical change. Needless to say, this in no way impinges on the empirical and theoretical comparison between the two theories.

7. Marxist theories

Marx held that technical change – the development of the productive forces – was the prime mover of history. His was not only an economic, but a technological conception of history. Man is a 'tool-making animal'.[1] 'It is not the articles made, but how they are made, and by what instruments, that enables us to distinguish different economic epochs.'[2] And the most authoritative statement of historical materialism – the Preface to the *Critique of Political Economy* – leaves us in no doubt about the absolutely central role of technical change in history. Moreover, the goal of history – the end state which teleologically justifies (and explains?)[3] the suffering of humanity – is a society in which the 'free, unobstructed, progressive and universal development of the forces of production'[4] finally becomes realized. There certainly is no other major social theorist who has attached a comparable importance to technical change. And since there is no other social theorist whose views have had a comparable impact, there is no need to justify the present chapter.

The chapter will deal mainly, but not exclusively, with Marx. There are not many later Marxist writings on technical change that have contributed to a theoretical advance, the main exception being the work of the 'labour process' theorists to be discussed below. I shall proceed as follows. First, I shall discuss what may be called, somewhat anachronistically, Marx's notion of the production function at the firm or industry level. (What is said here holds good exclusively for capitalism, since Marx does not discuss earlier modes of production in this fine-grained conceptual network.) The discussion may seem inordinately long, but I believe this to be justified by the lack (to my knowledge) of a good analysis of Marx's view on the topic. Secondly, I shall look at his explanation of the rate and the direction of technical change under capitalism. Here I shall have occasion to discuss the 'labour process view' referred to above. Thirdly, I shall briefly summarize his theory of the falling rate of profit and some of the controversies surrounding it. And finally I shall look at the broader historical theory of the interplay between the productive forces and the relations of production. Here again the exposition will be quite brief, since a fuller account is presented in Appendix 2.

a. Did Marx believe in fixed coefficients of production?

It is widely thought that Marx held the view that production (in industry)[5] took place with fixed coefficients of production.[6] I shall here argue that this interpretation is misleading or even wrong. To discuss this interpretation, we must first put it into a more precise form. Writing 'min $(x,y,z...)$' for the function that picks out the smallest of the numbers $x,y,z...$, a production function with *fixed coefficients* has the form

$$q = \min(a_1x_1, a_2x_2 .... a_nx_n) \tag{1}$$

where the x's are amounts of the various factors of production and the a's are constants. In other words, there are rigid technical complementarities between the factors of production, somewhat like those which exist between the pieces of a jigsaw puzzle. If you have two units of piece no. 78 in the puzzle and 1000 units of all the other pieces, then only two complete puzzles can be assembled, and the other pieces are all strictly useless. This implies (i) that any extra amount of any input (in an initial situation where $a_1x_1 = a_2x_2 .... = a_nx_n$) gives zero increase in output, and (ii) that a withdrawal of t % of any input (from the same initial situation) entails a reduction of t % in output. There is no 'mariginal product' and no possibility for substitution between the factors.

Fixed coefficients of production also imply constant returns to scale. There is no logical connection, however, between rigid complementarity and constant returns to scale. To see this, we may introduce the notion of *fixed proportions* of production.[7] We define this notion by retaining expression (1), while letting the coefficients of proportionality depend on the level of output:

$$a_i = f(q_i) \tag{2}$$

160 THEORIES OF TECHNICAL CHANGE

We also lay down the following condition:

$$\frac{f_i(q_o)}{f_n(q_o)} = \frac{f_i(q_1)}{f_n(q_1)} \quad . \text{ for all } i, q_o, q_1 \tag{3}$$

Condition (2) states that for any given level of output q, there is one and only one efficient input combination, i.e. exactly one combination in which there is no slack. That is to say, the factors of production have to be employed in certain definite amounts if one desires a certain output produced in a non-wasteful way. To these n amounts there also correspond *n-1* factor ratios. Condition (3) states that these ratios are the same for all levels of output. Take a numerical example, with $n=2$. Imagine that with 2 units of x_1 and 1 unit of x_2 we can efficiently produce 10 units of output. If we only impose condition (2), then it is conceivable that the efficient input combination required to produce 40 units is 6 units of x_1 and 4 units of x_2. If in addition we impose condition (3), this possibility is excluded. It would be compatible with conditions (2) and (3), however, that for the efficient production of 40 units, 6 units of x_1 and 3 of x_2 are required, for then the factor ratio is 2:1 at both levels of output, as required by condition (3). Observe that production in this case would not take place with constant returns to scale, although there would be fixed proportions of production. If, however, we impose condition (1) and take the a's as output-independent constants, then 40 units can only be produced efficiently with 8 units of x_1 and 4 of x_2, since now there must be constant returns to scale.

These notions can also be given a simple geometrical interpretation. That there are rigid complementarities in production means that all isoquants have the shape of right angles parallel to the axes. If we have fixed coefficients of production, then we need only the unit isoquant to represent the production possibilities. If we only require fixed proportions of production (with conditions (2) and (3) fulfilled), we may need more than the unit isoquant to represent the production function, but we know that the corners of the isoquants all lie on a straight line through the origin. If, however, we only impose condition (2), the isoquants will not in general have their corners on a straight line from the origin. I shall use the term *variably fixed coefficients* of production for the case where we impose condition (2), but not condition (3). And I shall argue that this

MARXIST THEORIES 161

notion provides a better (although still incomplete) approximation to Marx's theory than the usual assumption that he held a theory of fixed coefficients. Let us look at some of the more important texts involved. In one place Marx refers (ambiguously) to the strict complementarity between labour and the means of production:

> The quantity of means of production must suffice to absorb the amount of labour, to be transformed by it into products. If the means of production at hand were insufficient the excess labour at the disposal of the purchaser could not be utilised, his right to dispose of it would be futile. If there were more means of production than available labour, they would not be saturated with labour, would not be transformed into products.[8]

There is a slight ambiguity here, since Marx does not say in so many words that there is a strict proportionality which determines the saturation of the factors. Consider, however, the following passage:

> The division of labour, as carried out in Manufacture, not only simplifies and multiplies the qualitatively different parts of the social collective labourer, but also creates a fixed mathemetical relation or ratio which regulates the quantitative extent of those parts – i.e. the relative number of labourers or the relative size of the group of labourers, for each detail operation.[9]

This applies to the relation between workers in manufacture. A similar principle holds for the relation between machines in the industrial technology, or machinofacture:

> Just as in Manufacture, the direct cooperation of the detail labourers establishes a numerical proportion between the special groups, so in an organized system of machinery, where one detail is constantly kept employed by another, a fixed relation is established between their numbers, their size and their speed.[10]

And finally there is a passage in which the principle of complementarity is stated in full generality:

162 THEORIES OF TECHNICAL CHANGE

> [The technical composition of capital is a proportion that] rests on a technical basis, and must be regarded as given at a certain stage of development of the productive forces. A definite quantity of labour-power represented by a definite number of labourers is required to produce a definite quantity of products in, say, one day, and – what is self-evident – thereby to consume productively, i.e. to set in motion, a definite quantity of means of production, machinery, raw materials etc.[11]

Here Marx explicitly relates the fixed input amounts – the definite quantities – to the level of output, as in condition (2) above, leaving open the possibility that the proportions in which the factors are employed may vary with the level of output. I now cite two passages in which this is explicitly affirmed. First concerning machinery:

> In a large factory with one or two central motors the cost of these motors does not increase in the same ratio as their horse-power and, hence, their possible sphere of activity. The cost of the transmission equipment does not grow in the same ratio as the total number of working machines which it sets in motion. The frame of a machine does not become dearer in the same ratio as the mounting number of tools which it employs as its organs, etc.[12]

Second, more crucially, concerning the relation between machinery and labour:

> [The] concentration of labourers and their large-scale co-operation, saves constant capital. The same buildings, and heating and lighting appliances etc., cost relatively less for the large scale than for small-scale production. The same is true of power and machinery. Although their absolute value increases, it falls in comparison to the increasing extension of production and the magnitude of the variable capital or the quantity of labour-power set in motion.[13]

The emphasis here is on the fact that with large-scale production, unitary costs are reduced *and* factor proportions change (although being rigid at each level of production). In fact, the last passage also asserts that large-scale production is more labour-intensive than small-scale production.

This indicates that Marx is not here dealing with economies of scale due to technical progress, since this in his view was largely capital-intensive, as will be argued later. Rather, I think, he has in mind scale economies arising within a given technique.

Marx, then, denied that the factors of production could be substituted for each other within a given technique. In this he was clearly wrong. There often are within-technique possibilities of substitution, as in the electrolysis example given in the Introduction to Part II. Moreover, to assert this one does not have to believe in the full neoclassical story of smooth substitution along an isoquant. To refute Marx it suffices to cite cases in which a given level of output can be produced efficiently with two input combinations. Nor do I believe one can defend Marx by invoking satisficing or 'putty-clay' models. When Marx argued that firms knew only the practice they were currently using, he did so because he thought that at any given moment of time there was only one efficient technique in existence, not because it might not pay the firm to acquire knowledge about other techniques. And while it is true that the possibilities for substitution are much reduced *ex post,* compared to what one can do *ex ante* on the drawing board, I do not think one can read this distinction into Marx, nor that it would suffice to justify his denial of substitution. Surely there always are some possibilities of substitution *ex post,* e.g. between labour and raw materials. With high wages it may be optimal (i.e. profit-maximizing) to accept some waste of raw material rather than hire more workers to reduce the waste.

On the other hand Marx was fully aware of the possibility of between-technique substitution, as in the following passage:

In some branches of the woollen manufacture in England the employment of children has during recent years been considerably diminished, and in some cases has been entirely abolished. Why? Because the Factory Acts made two sets of children necessary, one working six hours, the other four, or each working five hours. But the parents refused to sell the 'half-timers' cheaper than the 'full-timers'. Hence the substitution of machinery for the 'half-timers' . . . The Yankees have invented a stone-breaking machine. The English do not make use of it, because the 'wretch' who does this work gets paid for such a small portion of his labour that machinery would increase the cost of production to the capitalist.[14]

164 THEORIES OF TECHNICAL CHANGE

Marx also quotes approvingly the following passage from John Barton:

The demand for labour depends on the increase of circulating and not of fixed capital. Were it true that the proportion between these two sorts of capital is the same at all times, and in all circumstances, then, indeed, it follows that the number of labourers employed is in proportion to the wealth of the state. But such a proposition has not the semblance of probability. As arts are cultivated, and civilisation is extended, fixed capital bears a larger and larger proportion to circulating capital. The amount of fixed capital employed in the production of a piece of British muslin is at least a hundred, probably a thousand times greater than that employed in a similar piece of Indian muslin.[15]

And, lastly, there is this insightful observation:

[It is the difference between the price of the machinery and the price of the labour-power replaced by that machinery] that determines the cost, to the capitalist, of producing a commodity, and, through the pressure of competition, influences his action. Hence the invention now-a-days of machines in England that are employed only in North America; just as in the sixteenth or seventeenth centuries machines were invented in Germany to be used only in Holland, and just as many a French invention of the eighteenth century was exploited in England alone. In the older countries, machinery, when employed in some branches of industry, creates such a redundancy of labour in other branches that in these latter the fall of wages below the value of labour-power impedes the use of machinery, and from the standpoint of the capitalist, whose profit comes, not from a diminution of the labour employed, but of the labour paid for, renders that use superfluous and often impossible.[16]

There is no doubt that Marx, in these passages, implies that the capitalist has a 'choice of techniques' and that he makes that choice on the basis of relative prices. Why, then, have so many authors – including myself[17] – stated without much evidence or argument that Marx denied the possibility of technical choice? I believe there are three main reasons.

First, the literature overwhelmingly neglects the distinction between technique and technology.[18] Since Marx explicitly denies the possibility of intra-technique choice and substitution, it is then all too natural to conclude that he denies the possibility of any kind of choice. In fact, it would appear that among the writers on technology Marx is among the very few who have seen the need for a three-tiered structure of technology, although he confused the matter by his denial of intra-technique choice. He did recognize, however, intra-technique variation, when he argued that a given technique does not consist of a single practice to be operated in various multiples depending on the level of output, but rather involves a change of practice when the scale of operation is increased. And, as I have just argued, he certainly recognized that there are inter-technique choices to be made.

The second reason why many writers have imputed to Marx a theory of fixed coefficients of production, excluding any kind of technical choice, is that without some such assumption the labour theory of value – in one of the many interpretations of that phrase[19] – fails to hold. Marx was strongly committed to the view that labour values are prior to prices, in the sense that we must know the values in order to determine the prices, but not vice versa. It can be shown that the first part of this statement is unequivocally false, on any assumption about the technology.[20] The 'not vice versa' part of the statement, however, is true in a model with fixed coefficients, but fails in a model with technical choice. This is so because choices between efficient techniques can only be made on the basis of factor prices, which means that one must first determine techniques and prices simultaneously, and only afterwards will one be able to deduce the labour values from the techniques. Moreover, there may be multiple price equilibria compatible with the given technical data and the given real-wage commodity bundle, and each of these equilibria may give rise to a different set of labour values for the commodities.[21] There are, therefore, strong theoretical pressures within the Marxian system towards an assumption of fixed coefficients.

The third reason has to do with the reluctance of many Marxists to invoke rational-choice explanations. There is a widespread, if diffuse, belief that such explanations somehow involve a subjective element that is incompatible with Marxist materialism.[22] I believe that there is no such incompatibility; and even had there been, so much the worse for Marxism. There is no need to argue the problem in much detail, since Marx by his own practice shows that he is in no way averse to rational-choice explanation in economics. The labour theory of value has as a fundamental postulate the equalization of the rates of profit in all sectors, which comes about only because capitalists move their capital to the industry that at any given time has the highest rate of profit.[23] This, surely, is maximizing behaviour, not satisficing. Nor is capitalist behaviour compulsive, in the sense that workers' consumption behaviour in early capitalism can be said to have been compulsive. If the budget constraint and the calorie constraint together are strong enough to narrow down the feasible set of options to a single consumption bundle, then all talk about rational choice becomes redundant and easily ideological. But capitalist entrepreneurs did have genuine choices to make, concerning the level of output and the combination of inputs.

b. The rate and direction of technical change

I now turn to the 'micro-motives and macro-behaviour' involved in the Marxist theory of technical change under capitalism. As in earlier chapters, we are concerned with the explanation of the *rate* and the *direction* of technical change.

The classical Marxist explanation of the rate of innovation is very simple: capitalists innovate because they are forced to do so by competition, and they are able to innovate because they can draw on a stock of inventions, i.e. on science. Since Marx has little to say about science,[24] I shall deal only with the first issue. Capitalism is a dynamic system, in the sense that entrepreneurs tend to reinvest part of their profits in new production. This does not follow from the facts of private property in the means of production and of wage labour, since the capitalist could conceivably use his profits for hoarding or for luxury consumption. Marx, however, postulates that this will not be the case. The postulate can be looked upon as psychological or institutional, depending on the period with which one is concerned. In an initial stage there must have been some psychological drive that made the entrepreneur thirst for ever-increasing amounts of surplus-value, but later on he had no choice but to reinvest if he were to survive in the competition. (This distinction is made explicitly only by Weber,[25] but I believe it also makes sense of what Marx says on similar topics.)

In an early stage of capitalism, reinvestment could take place without technical change, as long as there were pre-capitalist sectors in which to invest and expand. Once, however, the capitalist mode of production permeates the whole economy – and disregarding imperialist and colonial ventures – reinvestment cannot take place indefinitely on a constant technical basis. Reading a bit of Schumpeter into Marx, we can observe that this is so because the demand for the existing consumer goods ultimately becomes saturated, at least at the ruling prices. Also, staying closer to Marx himself, there will ultimately develop a scarcity of labour, at least if (and as long as) the rate of growth of the labour force is below that of the rate of reinvestment. Hence, saturation in the product market and pressure in the labour market will force the capitalist to innovate, both in order to be able to sell his goods at a profitable price and in order to cut his costs. One will observe product innovation as well as process innovation, since the saturation can be overcome both by producing new types of commodities and by producing the old ones more cheaply. I repeat that there is little in Marx about innovating being induced by demand saturation, but it fits in well with the kind of story he is telling. By contrast, it is possible that Marx did see rising wages as an inducement to innovate, although we shall have to take a closer look at his argument.

There is, then, a relentless pressure towards innovation. The question, for the individual capitalist and for the capitalist system, is: innovate or perish. The individual capitalist confronts a situation in which all others innovate, and so he is damned if he doesn't. The system confronts the situation that if there is little innovation, saturation of demand and rising wages will squeeze the profit rate below the minimum that capitalists are willing to accept. As we shall see below, Marx argued that even with innovation – and because of it – the profit rate will tend to fall. The system, in fact, is damned if it does and damned if it doesn't. Postponing this question for the moment, let me just observe that it is not the system's need for innovation that forces the individual capitalist to innovate. To assume so would be to fall victim to teleological thinking and functional explanation. The individual entrepreneur invests and innovates because it is rational for profit maximization or necessary for survival.

These are very broad and general arguments, which cannot tell us much about the rate of technical change that will actually be observed. Marx offers, however, one important analytical argument that bears directly on the rate of technical change under capitalism, compared to that which would have been observed in a communist society. In a passage quoted above Marx asserts that the capitalist profit comes 'not from a diminution of the labour employed, but of the labour paid for'.

And he adds in a footnote: 'Hence in a communistic society there would be a very different scope for the employment of machinery than there can be in a bourgeois society'.[26] The point is that in certain cases new techniques that would be superior from the point of view of labour-minimizing will not be so from the point of view of profit-maximizing. For a full discussion of the relation between 'viable' or cost-reducing and 'progressive' or labour-reducing technical change, the reader is referred to the recent book by John Roemer.[27] Here I shall illustrate the idea by a simple example, which will also be useful in explaining some aspects of the 'capital controversy' referred to in earlier chapters.

Assume that we are dealing with an economy that produces one final (consumption) good. This can be done by using Practice I or Practice II, each of which involves a particular kind of intermediate (capital) good. Let us write out for Practice I the equations that determine the price ratio and the rate of profit, assuming that the technical coefficients and the real wage are given. In order to produce one unit of the capital good one needs a_{11} units of the capital good and a_{12} units of labour. In order to produce one unit of the consumption good one needs a_{21} units capital and a_{22} units labour. We set the price of the capital good equal to 1 by convention, since we are interested only in relative prices. The real wage (in units of the consumption good) is w, the rate of profit is r, and the price of the consumption good is p. We assume – contrary to Marx, but with little loss of generality for the present purpose – that the capitalist does not calculate profits on the advance for wages. We can then lay down the following equilibrium conditions:

$$a_{11}(1+r) + a_{12} \cdot w \cdot p = 1$$
$$a_{21}(1+r) + a_{22} \cdot w \cdot p = p$$

Each equation says that wages + capital + profit on capital must equal price. From these equations we can determine r and p in terms of the technical coefficients. In particular, we can obtain an explicit analytical expression of r as a function of w. In a coordinate system with w measured along the horizontal axis and r along the vertical, this relation will have a downward-sloping graph, which may or may not be a straight line. (The significance of this will be made clear later.) In Fig. 5 I have drawn two such curves, shown as straight lines for simplicity. At any wage rate below w_o, Practice I is the most profitable and will be chosen by an entrepreneur

Fig. 5

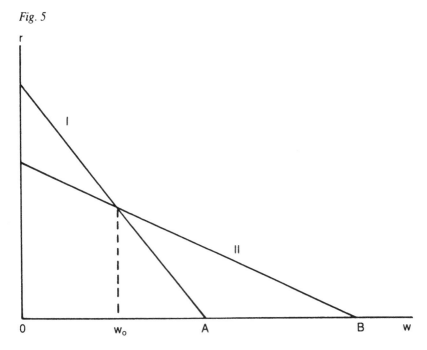

out to maximize the rate of profits on capital. At wages above w_o Practice II is preferred. Observe now that on labour- minimizing grounds Practice II is unambiguously superior. This is so because the distance OA (respectively OB) can be interpreted as the amount of production per worker when Practice I (respectively II) is employed: at all points on the horizontal axis the profit is zero because the whole net product goes to wages. Since Practice II allows more output per worker, it also has lower labour time per unit of output and in that sense is more 'progressive' than Practice I. Nevertheless, at low wages the socially inferior Practice I will be preferred by a profit-maximizing capitalist.

As a brief aside, it may be pointed out that the linearity of the w-r curves is a sufficient condition for the notion of aggregate capital to make sense in this simple model. If at least one of the curves is non-linear in shape, there may be multiple points of intersection between the curves, so that one practice may be profit-maximizing at low wages, the other at intermediate wage levels, and the first again at high wages. This, in turn, is incompatible with the neoclassical notion that an individual practice corresponds to a single point on an isoquant that is convex towards the origin (as in Fig. 1 and 2 of Ch. 4 above). As mentioned earlier, it is not clear how often such 'reswitching' will occur in empirical cases, but surely one cannot be content with hoping that it will not arise.

Going back to the main argument, Marx certainly was correct in thinking that communism, other things being equal, would be superior to capitalism in introducing machinery to save on human drudgery. This, however, does not resolve the Schumpeterian issue whether other things will in fact be equal. It might turn out that this dynamic suboptimality of capitalism (coming in addition to the static suboptimality introduced by the patent system) cannot be eliminated without fatally reducing the motivation to innovate. The superiority of capitalism with respect to the *search* for new techniques might more than offset, that is, the inferiority with respect to the *selection* of socially useful innovations. Marx, however, believed that communism would be superior in both respects. As is more fully explained in Appendix 2 below, he believed that the intensity of innovative effort would be vastly greater under communism than under capitalism.

The above argument was put forward on the assumption of a given and constant wage rate. For some wages, e.g. wages superior to w_o in Fig. 5, progressive innovations will be rejected by capitalist entrepreneurs because they are not viable, i.e. do not reduce costs. A slightly different argument can be constructed by assuming that some innovations will bring about a change in the wage rate, and that the capitalists may evaluate them by anticipating this change. We must remind ourselves that innovations have to be *embodied* in tools, machinery and factory design, and that the latter in turn may shape and influence working-class consciousness and combativity. If a given innovation is viable at the pre-innovation wage rate, but not at the wage rate that will obtain when the machinery embodying it has been introduced, an intelligent capitalist might abstain from adopting it. Conversely, there is the logical possibility that a capitalist might introduce techniques that *ex ante* are non-viable, if *ex post* they fetter the wage struggle of the workers. These hypotheses form the core of the 'labour process approach' to technical change. Compared to the theories studied in previous chapters, they involve a radical change of perspective. The labour force is not merely regarded as a 'factor of production' whose price may be subject to fluctuations, but as a self-conscious class actively opposing and frustrating the extraction of surplus-value. The capitalist also appears in a different role. We have seen him successively as the rational calculating machine of neoclassical economics, as the 'unbound Prometheus' of Schumpeter and his followers,[28] as the satisficer who innovates only when compelled to by circumstances, and as the myopic tinker who progresses only through 'learning by doing'. Now it appears that the entrepreneur confronts not only the forces of nature and the equally impersonal forces of the market, but that he also has to deal directly with men – with whom he engages in a game of mixed conflict and cooperation.[29]

The origin of this view is to be found in these passages:

> In England, strikes have regularly given rise to the invention and application of new machines. Machines were, it may be said, the weapon employed by the capitalists to quell the revolt of specialized labour. The *self-acting mule,* the greatest invention of modern industry, put out of action the spinners who were in revolt. If combinations and strikes had no other effect than that of making the efforts of mechanical genius react against them, they would still exercise an immense influence on the development of industry.[30]
>
> But machinery acts not only as a competitor who gets the better of the workman, and is constantly on the point of making him superfluous. It is also a power inimical to him, and as such capital proclaims it from the roof tops and as such makes use of it. It is the most powerful weapon for repressing strikes, those periodical revolts of the working class against the autocracy of capital. According to Gaskell, the steam engine was from the very first an antagonist of human power, an antagonist that enabled the capitalist to tread under foot the growing claims of the workmen, who threatened the newly born factory system with a crisis. It would be possible to write quite a history of the inventions made since 1830, for the sole purpose of supplying capital with weapons against the revolts of the working class. At the head of these in importance stands the self-acting mule, because it opened up a new epoch in the automatic system.[31]

This line of argument has been extensively developed during the last decade or so. An influential but non-analytical work was Harry Braverman's *Labor and Monopoly Capital.*[32] More recently, special issues have been devoted to the problem by *Monthly Review* (1976), *Politics and Society* (1978), and *Cambridge Journal of Economics* (1979). I shall limit my discussion, however, to three theoretical studies by Amid Bhaduri, Stephen Marglin, and John Roemer.

Bhaduri's work on agriculture in the Indian sub-continent shows how landowners might resist an apparently profitable innovation that could undermine their extra-economic hold on the tenants.[33] In this semi-feudal sharecropping economy landowners' revenue is made up of two parts: their share of the crop and the interest on the consumption loans accorded to the tenants. The interest rate typically is very high, since loans are given in rice evaluated at the very high pre-harvest prices and paid back in rice evaluated at low post-harvest prices. The effect of a given innovation on the income of the landowner will, therefore, have two components. In the first place the increase in productivity will increase the size of the share that goes to the landowner. In the second place, however, it will also increase the size of the tenant's share, and enable him to reduce and perhaps eliminate altogether his debt to the landowner, reducing the interest income of the latter. It goes without saying, therefore, that the landowner will only accept the technical changes that do not reduce the sum total of his income, i.e. whose net effect on his income is positive. Bhaduri also argues, however, that the landowner will resist any innovation that enables the tenant to free himself from the debt bondage, even if it should increase the size of the landowner's share so much that the net effect on his income is positive. This is so because the debt bondage is a means of keeping the tenant's percentage share as low as it is, and in fact an essential condition for the social power of the landowner. The landowner, in other words, is not only concerned with the net effect of innovations on his income, but also with the long-term effect.

Closer to Marx's own argument is an article by Stephen Marglin, subtitled 'The origins and functions of hierarchy in capitalist production'. The phrase 'origin and function' indicates the ambiguous character of Marglin's work, hovering between intentional and functional explanation in a way that is never made completely clear. His argument is that the division of labour characteristic of early capitalism and the factory system of mature capitalism were not introduced because of their superior technical efficiency. Rather the division of labour was introduced to deprive the workers of control over the *product* of their labour, and the factory then deprived them also of control over the work *process*. The 'social function' of the hierarchy inherent in manufacture and machinofacture is not technical efficiency, but capital accumulation. New methods are introduced because they are profit-maximizing, not because they are technically superior. True, the factory system, once introduced, proved to be a vehicle for innovation, but it was not introduced for this reason, nor was it in any way indispensable (technologically) for innovation.

Marglin's article is conceptually acute and offers some interesting historical evidence. Yet it appears to me seriously confused on the level of explanation. Consider first his argument that the division of labour was an instance of 'divide and conquer'. It is not all clear whether he does not really have in mind the 'tertius gaudens' effect briefly mentioned in Ch. 2 above, i.e. the ability of the capitalist to exploit a state of isolation among the workers that he did not have to create in the first place. In fact, Marglin explains very well why the individual worker was not able to integrate his own work: he did not have the necessary capital 'to set aside for the mistakes inherent in the learning process'.[34] Marglin may be right when arguing that there were no technical obstacles to the individual worker performing in series the operations that capitalist manufacture performs in parallel, but this does not show that the goal or the function of capitalist enterprise was to provide an institutional obstacle. Nor is it easy to see how it could have done so.

Moreover, Marglin confuses the reader by choosing baselines for comparison that are sometimes actual, sometimes counterfactual. His declared aim is to explain the rise of manufacture and machinofacture by their exploitative efficiency while relegating technical improvement to a secondary place, thus reversing the traditional order of priorities. This argument logically requires confirming instances of the following kind: institution A wins out over institution B because it is better at exploiting the workers, without any concomitant technical superiority. In the case of the transition to the factory system, Marglin provides an instance of this kind, when pointing to the success of the Arkwright factory compared to the technologically equivalent Wyatt-Paul enterprise that failed to 'subdue the refractory tempers of work-people'.[35] When discussing the introduction of division of labour in manufacture, Marglin offers no such instances. Here he only points to the technological equivalence of a counterfactual system of manufacture owned and controlled by the workers. Clearly this is irrelevant for explanatory purposes. For these the central fact is that the capitalist division of labour was more efficient than the system it replaced, not that it was more exploitative than the counterfactual system. The latter fact would have been relevant if capitalism had been instrumental in preventing the counterfactual system from becoming actual, but, as argued above, Marglin does not provide any grounds for this view.

Finally Marglin has a somewhat cavalier attitude to the evidence that would be needed to show that the workers' alienation from their work and their product is a consequence of the capitalist system which also *explains* the rise of that system. Unlike most other writers of this school, he explicitly confronts the problem of evidence:

> Hard evidence that 'divide and conquer' rather than efficiency was at the root of the capitalist division of labour is, naturally enough, not easy to come by. One cannot really expect the capitalist, or anybody else with an interest in preserving hierarchy and authority, to proclaim publicly that production was organized to exploit the worker. And the worker who was sufficiently acute to appreciate this could, in the relatively mobile societies in which the industrial revolution first took root, join the rank of the exploiters.[36]

In other words, little evidence is left because at the time there was nobody who both understood what was happening and had an interest in denouncing it. A worker smart enough to recognize that he was exploited would – because he was so smart – become an exploiter, and so presumably would not complain about the system. Nor would those who already were exploiters in the first place. This is, in my view, a perfect example of the immunizing strategies that have led Marxist social science into near-universal disrepute. I am not saying that Marglin is wrong. I am only objecting to the use of this unsubstantiated assertion as a blanket excuse for dispensing with evidence about mechanisms and intentions. In all fairness it should be added that Marglin clearly is concerned with offering evidence, or 'examples'. Yet I believe that his attitude towards such evidence is systematically ambiguous, because of a semi-functionalist, semi-conspiratorial attitude that enables him to explain phenomena in terms of their consequences without stooping to specify a mechanism.

John Roemer has tried to incorporate the labour process approach in a formal model of a capitalist ecomomy. In my view he fails, yet interestingly so. Roemer first defines the notion of a *reproducible solution* for a capitalist economy, in which workers have a given subsistence bundle and capitalists maximize profits, faced with a 'production set' that allows for choice of techniques. In a reproducible solution net output should at least replace workers' total consumption; intermediate outputs and workers' consumption must be available from current stocks, and prices are such as to allow input markets to clear. He then proves a generalized version of what Morishima has called the 'Fundamental Marxian theorem'. Given the technical and behavioural assumptions briefly indicated above, the following statements are all equivalent:

(i) There exists a set of productive activities which – given the wage bundle – yield a positive rate of exploitation.
(ii) There exists a reproducible solution yielding positive total profits.
(iii) All reproducible solutions yield positive total profits.
(iv) All reproducible solutions yield positive rates of exploitation.[37]

Next, Roemer tries to extend this model by letting workers' consumption be socially determined, i.e. be an endogenous element of the model. He postulates that workers' consumption is a continuous function of the technology chosen by the capitalists. The argument of this function is a set of n vectors, each of which represents the inputs and outputs of a particular productive acitivity. In other words, the (common) wage bundle consumed by the workers is a continuous function of the economy-wide technological structure. Given this assumption, Roemer proves first the existence of reproducible solutions with endogenous wages and then a weaker form of the fundamental Marxian theorem. The main interest lies in the weakening of the theorem under the new assumption. Roemer provides one example to show that reproducible solutions with positive profits and with zero profits may coexist, i.e. 'there may be different equilibria for the economy in which the relative balance of class forces is different, because of the effect the choice of technique has on workers' organization'.[38] Another example shows that

> It may be possible for capitalists to choose techniques at which there is exploitation, but if they insist on *maximizing* profits at given prices, they must choose techniques that will alter the balance of class forces to such an extent that subsistence requirements become too great for

reproducibility. Exploitation is possible if capitalists limit themselves to suboptimal techniques: but if capitalists choose profit-maximizing techniques, the subsistence wage will be driven up, eliminating profits.[39]

Clearly these are interesting possibilities, yet somewhat remote from actual capitalist economies. In particular, the fact that the profit of each capitalist depends on the technology chosen by all makes it hard to see how – in the absence of an improbable form of collective action – the capitalists could limit themselves to suboptimal techniques in order to maximize profits. This, however, is not my main objection to this model. Rather I find it odd that working-class combativity should depend on the input-output coefficients, and even odder that it should vary continuously with these coefficients, as is required for the existence theorem to go through. True, working-class consciousness is shaped in part by the structure of the labour process. Equally true, labour processes with different structures will typically have different input-output coefficients. But there is no reason to believe that the impact of the labour process on class consciousness is mediated by the input-output coefficients. And, as far as I can understand, this would be required for the continuity assumption to make sense. Although I cannot offer a rigorous disproof of Roemer's view that the assumption of continuity is not very harsh 'considering the social process that underlies [the assumption of endogenous wages]',[40] I strongly believe that class consciousness is shaped by qualitative features of the labour process that cannot (at present) be captured in quantitative terms.

I now turn to the *direction* of technical change under capitalism. It might appear that the labour process approach would also be relevant for this issue, by postulating a connection between working-class combativity, high wages, and labour-saving innovations. Marx, in a passage quoted above, states in fact that machines were 'the weapon employed by the capitalists to quell the revolt of specialized labour'. Although this mechanism certainly operates, I do not believe it to be universally or predominantly present. Mechanization, in fact, may go together with fewer and more highly skilled workers who are also more intractable and ungovernable. In early industrialization certain labour-saving innovations may have implied less need for skilled labour, but there is no reason why this should be the case generally. Given two feasible innovations that are equally cost-reducing at the pre-innovation factor-price ratio, there is no reason to think that the one that has a labour-saving bias should also *at the micro-level* change the class forces so as to bring about a lower wage rate than the other – nor to assume a trend in the opposite direction. Again I believe that the qualitative factors are too subtle to be modelled in a simple quantitative way.

Marx certainly believed that labour-saving innovations, in the form of a 'rising organic composition of capital', dominated the development of capitalism and – as explained below – would ultimately bring about its downfall. True, he knew very well that technical change was not always labour-saving. There is an interesting chapter in *Capital III* on 'Economy in the employment of constant capital', which explains why intra-technique economies of scale can be capital-saving. On the other hand, I do not believe he ever mentions capital-saving change *of* technique. We do not have to go to such recent examples as explosives or the wireless for examples of capital-saving innovations in this stronger sense. Examples were there to be seen before Marx,[41] but he did not see them. The dramatic increases of productivity associated with changes of – as opposed to within – techniques were all, on his view, labour-saving.

Yet it is hard to reconstruct, on the basis of Marx's writings, an argument for the labour-saving bias. It is tempting to read into Marx a Hicksian argument for the predominance of labour-saving innovations, as is done by Maurice Dobb (with an explicit reference to Hicks) and Paul Sweezy.[42] Sweezy justifies his interpretation by citing a passage from *Capital I* in which Marx, indeed, explains mechanization as induced by rising wages.[43] Yet the passage in my view is marginal, dealing with agriculture and not with industry. It *is* evidence for Sweezy's view, but in my view it is outweighed by the fact that in the crucial chapters of *Capital III* on the falling rate of profit there is no hint at a link between rising wages and mechanization.

In these chapters Marx seems to be under the spell that innovations simply *must* be labour-saving, for how otherwise could a slowly growing number of workers set in motion a rapidly growing mass of raw material and create a rapidly growing mass of products? A tentative reconstruction of his argument is the following.[44] (i) Economic growth implies or is synonymous with more output per worker. (ii) If more is to be produced per worker, each worker must handle more raw material. (iii) In order to be able to handle more raw material, the worker needs the help of machinery. (iv) Since constant capital mainly consists of raw material and machinery, it follows that the amount of constant capital per worker must increase. Premises (ii) and (iii) appear compelling, but in fact are not. They embody a narrow vision of technical change that excludes such dramatic capital-saving innovations as explosives, which enable the worker to handle vastly more raw material using vastly less capital. I conclude that Marx has no valid argument for the dominance of labour-saving innovations. The Hicksian argument, if it could successfully be imputed to Marx, was shown in Ch. 4 above to be invalid, and the argument set out above is based mainly on pre-theoretical conceptions.

c. The falling rate of profit

The theory of the falling rate of profit in Marx is somewhat peripheral to the concerns of the present work, dealing with the consequences rather than the causes of technical change. Yet a brief discussion is included here, to show in which sense Marx believed that technical change would bring about the downfall of capitalism. This will enable us in the next section to compare the theory of the falling rate of profit with the more general theory that *all* modes of production wax and wane as a function of technical change. Since I can refer the reader to the exhaustive discussion by John Roemer,[45] I can limit myself to a few general remarks.

All the classical economists believed there was a tendency for the rate of profit to fall. Marx, taking over their conclusion, argued for it in a radically different way. They found the source of the falling rate of profit in diminishing returns in agriculture. More workers require more food, which can only be made available on less fertile land or through more intensive work on the given land, both resulting in higher food prices, rising wages and falling profits. Marx, however, found the source of a falling rate of profit in industry within industry itself. In *Capital III* the section on the falling rate of profit precedes the discussion of ground rent, whereas both Ricardo and Malthus had the treatment of agriculture precede that of profits in their main works. Marx remarked that his predecessors explained the falling rate of profit by invoking organic chemistry, while he sought the explanation in the contradictions of capitalism as a mode of production.[46] Also, whereas the classical economists saw technical change as a force counteracting the tendency of the rate of profit to fall, Marx argued that the fall occurred as a result of labour-saving technical change. Today it is clear enough that they were right and he was wrong, but it remains to be seen why he arrived at these counter-intuitive views.

Marx assumed that the average (economy-wide) rate of profit was $r = \frac{S}{C+V}$ where S is the total surplus-value realized in the economy, C is the value of the constant capital, and V the value of the variable capital. (Value here means the labour time required to produce the various commodities.) This assumption is not justified, because of the so-called 'transformation problem', i.e. the fact that prices typically will not be proportional to values. For simplicity, however, I shall ignore this difficulty and accept Marx's formula for the profit rate.[47] I shall also assume that the constant capital is completely used up during the production process, i.e. that it all consists of circulating capital. Contrary to a recent argument, this simplification makes no qualitative difference to the conclusions of the analysis.[48] If in the profit rate formula we divide through by V, we can represent the rate of profit as $r = \frac{S/V}{C/V + 1}$. Here the numerator S/V is the *rate of surplus-value* or the rate of exploitation. C/V in the denominator is the *organic composition of capital*, an expression of the amount of capital employed per unit of variable capital. Unlike the concept of 'capital intensity', it is not a purely technical relationship, since it may be modified by a change in wages alone. If, however, we keep the wage rate constant, the organic composition of capital will reflect purely technical relations.

Marx's argument goes as follows. In the course of capitalist development, there is a predominance of labour-saving innovations, which lead to an increase in the amount of machinery (and other constant capital) employed per worker and hence to a rising organic composition of capital. Assuming a constant rate of surplus-value, this will lead to a fall in the rate of profit. If, moreover, we impute to Marx the Hicksian argument for the predominance of labour-saving inventions, we have gone full circle. Initially, there is a pressure on the labour market bringing about higher wages and falling profits. A rational entrepreneur, Hicks (wrongly) argued, will then react with a labour-saving bias in his search for innovations. And indeed, this would be a good idea for the individual entrepreneur were he the only one to have it. But when all capitalists have it simultaneously, the collective consequence is the very opposite of what they individually desired and intended. When all firms save on labour, the total amount of living labour employed in the economy must fall. But that way lies disaster, since living labour is the only source of surplus-value and thus of profits. An initial fall in the rate of profit will induce individuals to undertake actions whose collective consequence is to accelerate the fall in the rate of profit.

This attractive argument unfortunately has no validity whatsoever. Basically it fails because Marx over-emphasizes the labour-saving side of the innovations and under-emphasizes the fact that they also increase productivity. This mistake shows up in two different ways, pertaining respectively to the numerator and the denominator in the formula for the rate of profit. First, technical change will lead to a cheapening of the consumption goods. This is compatible with a constant rate of surplus-value only if we assume an increase in real wages. But this is a strange assumption to make, at least within a Hicksian framework, since 'the net effect of all the capitalists' behaving in this way . . . is to create unemployment which in turn acts upon the wage level'.[49] Independently of this Hicksian argument, a constant rate of surplus-value under conditions of rising productivity is hard to reconcile with Marx's view that the income of the workers was falling, if not in absolute terms at least relative to other incomes.[50]

Secondly, technical change will also lead to a cheapening of the capital goods, thus bringing about a fall in the value of constant capital C and hence in the ratio C/V. To bring out this more clearly, we may envisage the following scenario. Initially there occurs technical change in all industries, which is labour-saving when capital goods are evaluated at pre-innovation values (or prices). We may think of this as an increase in the 'number of machines per worker', a crude but useful way of visualizing the change. The innovations will lead to new input-output coefficients and hence to new equilibrium values (and prices). It is then logically possible for the new techniques to be *capital-saving* in all industries when the capital goods are evaluated in the new values (or prices).[51] This will occur if the productivity increase is so important that the fall in the value (or price) of each machine more than offsets the increase in the number of machines per worker. It appears clear that Marx believed in the predominance of labour-saving inventions in the *ex ante* sense, but failed to recognize that this bias does not necessarily carry over to the *ex post* situation.

I am not saying that this scenario must occur, only that it can occur. To refute the theory that the rate of profit must fall, it suffices to demonstrate the possibility of one instance – compatible with all of Marx's assumptions – where it will not. And it is of no avail to argue that the law of the falling rate of profit must be understood as a 'tendency law', which can be disturbed by such 'counteracting forces' as a cheapening of the elements of constant capital, foreign trade etc. As observed by Mark Blaug, there is a proper and an improper use of the notion of 'tendency laws' in classical economics.[52] Properly understood, a tendency law is a law that holds under certain ideal conditions. The law of falling bodies is a tendency law in this sense, because it assumes that there is no air resistance. Similarly one may abstract from foreign trade in an analysis of the trend of the profit rate, and possibly arrive at a valid tendency law in this sense. However, the notion is improperly used when one disregards the cheapening of the elements of constant capital, since this 'counter-tendency' stems from the same process as also generates the 'main tendency'. We want to know about the *net effect* of technical change on the rate of profit, and the decomposition of this into a main tendency and a countertendency is artificial and of little interest. I conclude that the notion of a tendency law in this latter, improper, sense belongs to the arsenal of immunizing stratagems of Marxism, along with such phrases as 'relative autonomy' and 'determination in the long run', or the argument invoked by Marglin to explain the paucity of evidence for his theory.

d. The development of the productive forces

In Appendix 2 below I discuss at some length Marx's theory of productive forces and relations of production. Here I only want to focus on some narrow explanatory issues that are raised by the theory. In particular, I want to show how G. A. Cohen, by resolving a long-standing problem in the interpretation of historical materialism, has also made it vulnerable in new ways.

Marx was committed to two – apparently incompatible – theses concerning the relation between the productive forces and the relations of production. On the one hand, he believed that the productive forces 'determine' or 'condition' the relations of production. On the other hand, he repeatedly asserts that the relations of production have a decisive effect on the forces, being successively 'forms of development' and 'fetters' on technical change. The puzzle is how entity A can 'determine' or 'have primacy over' entity B when at the same time B is said to exercise a crucial influence on A. (The problem arises also when for A we take the relations of production and for B the political and legal superstructure.)[53]

Marxists have tended to evade this issue. Some have talked vaguely about 'determination in the long run', others have in effect abandoned the theory by referring to the 'dialectical interaction' between forces and relations of production.[54] G. A. Cohen has shown, however, that it is possible to reconcile the two theses by explaining the relations of production in terms of their effect on the forces of production.[55] The relations of production at any given time are what they are because of their ability to promote the development of the productive forces, and they change when they no longer have this ability. Thus the relations of production at any given time have a *causal* primacy over the productive forces, and the latter an *explanatory* primacy over the former.

Exegetically, I believe this argument makes good sense. The historical task of capitalism according to Marx was to develop the productive forces; it is there becuse it develops them and will disappear when it no longer does so (optimally). Similarly, at least in principle, for precapitalist modes of production. To Cohen's exegesis one should add, however, an argument offered by Philippe van Parijs.[56] He points out that the forces of production are *doubly* involved in the explanation of the relations of production. (1) It is the level of the productive forces which determines what relations of production are optimal. (2) The relations of production are what they are because they are optimal for the development of the productive forces. Logically speaking, these are distinct views; one could hold one of them but not the other. A theory embodying the first view, but not the second would hold that the relations of production are what they are because they are optimal for something other than the development of the productive forces, but that the level of the productive forces is what determines which relations are optimal for that purpose. A theory embodying the second view but not the first would hold that the relations are what are because of their impact on the forces, but that there is something else which determines what relations are optimal in that respect. Van Parijs suggests that the 'something else' in the first case might be repression of the people, and in the second case location along the centre-periphery dimension. It is clear that Marx was committed to both (1) and (2) above, and that his argument therefore involves both a causal and a teleological element. In what follows, however, I shall focus only on the latter, i.e. the relation (2) above.

The question then becomes whether Marx was *correct* in believing that the relations of production could be explained in terms of their effect on the productive forces. Cohen suggests that we may justify his explanation by a consequence law, to the effect that whenever a new set of relations would bring about an accelerated development of the forces, a new set appears on the historical scene.[57] But it is made clear in Appendix 2 that the number of instances backing that law is pitifully small, perhaps dwindling to just one. Capitalism did indeed bring about a dramatic revolution in technical progress, but it would be foolhardy to use this case as sole evidence for the general law that is invoked to explain it. In addition, of course, there are the general problems involved in Cohen's account of functional explanation. These were discussed in Ch. 2 above, and I shall not repeat my objections here.

An alternative would be to prove Marx right by pointing out the specific mechanism whereby the productive forces 'select' the relations most able to further their development. Cohen briefly sketches one possible mechanism,[58] but the argument is clearly unsatisfactory.[59] A more plausible suggestion could be the following. The rise of new relations of production may well be largely accidental, i.e. the outcome of a complex and opaque class struggle not easily subsumable under a general scheme. The rise of capitalist relations in English agriculture – as opposed to the emergence of a free peasantry in France – provides an example.[60] To explain this divergent development we must invoke minute details of centuries of class struggle. There is no general scheme capable of explaining the emergence of capitalist relations in England through the 'need' for development of the productive forces. Once these relations had emerged, however, they soon became dominant in other countries as well, through their sheer economic and technical superiority. One can explain functionally the diffusion of new productive relations, but not their emergence. A similar but more hypothetical argument is sketched in Appendix 2 for the emergence and diffusion of communism.

This argument, if accepted, sustains a weak functional explanation of the relations of production. It could be strengthened were one able to show that in the course of pre-capitalist development capitalist relations of production *had* to arise somewhere, sometime, just as in the theory of natural evolution one may assume that any point mutation eventually will arise in some organism. With this added assumption, one could argue not only that capitalism exists because it is better at promoting the development of the productive forces than the previous arrangement, but also that the previous arrangement had to disappear because it was no longer optimal. But I fail to see how the inevitability of the 'capitalist mutation' could be demonstrated. This means that the explanation considered here differs from that offered by Cohen in his book in two respects. First, the superiority of capitalism explains its dominance, not its emergence. Secondly, the dominance of capitalism is not a necessary phase in world history.

I conclude with an unresolved puzzle. There appear to be in Marx two theories of the downfall of capitalism. On the one hand, there is the theory of the falling rate of profit, stating that capitalism will disappear because labour-saving innovations will bring accumulation to a halt by reducing the source of surplus-value and of profits. On the other hand, there is the theory of productive forces and relations of production, stating that capitalism – like any other mode of production – will disappear when and because it is no longer optimal for the development of the productive forces. Both theories invoke technical change in crucial, yet very different, ways. Unless one is prepared to argue that the downfall of capitalism according to Marx was overdetermined, in the sense that two distinct sufficient causes were operating, this dual explanation points to a serious inconsistency in Marx's thought.

APPENDIX 1
Risk, uncertainty and nuclear power

I. Introduction

The main purpose of this paper is to apply decision theory to illuminating the structure of the choice between alternative modes of energy production that most Western societies face at the present time.[1] A secondary purpose is to use the energy problem as an illustrative introduction to decision theory for readers not acquainted with this discipline. The discussion is throughout normative, except for the final section, in which I discuss some of the reasons why actual decision processes may not conform to what I see as the rational procedure.

I do not attempt to answer the substantial question: what *will* happen if we choose one or another of the proposed energy forms? Rather I am arguing that for important parts of the energy issue this question cannot be answered: and that *this impossibility is the substantial result which must be the basis for choice.* The operative notions here are the ones of *risk and uncertainty,* two forms of ignorance that differ profoundly in their implications for action. Decisions under risk are present when we can assign numerical probabilities to the various answers to the question 'What will happen?' Decisions under uncertainty imply that we can at most list the possible answers,[2] not estimate their probabilities. To this analytical distinction there corresponds a distinction between two criteria for rational choice, which may roughly be expressed as 'maximize expected utility' and 'maximize minimal utility'. I argue that there is at least one issue in the energy choice – the proliferation of nuclear weapons – that constrains the question into one of decision under uncertainty, with the implied imperative to act as if the worst that can happen, will happen.

II. Decision theory

Decision theory deals with the problem of choosing one course of action among several possible courses. The normative branch of the theory, with which I am here mostly concerned, deals with the rational way of making this choice.[3] The empirical branch studies how actual decisions are made.[4] A further subdivision, cutting across the first distinction, is between individual and collective choice. Within the theory of rational individual choice some further distinctions are made, pertaining to the degree of certainty attached to the choice variables.

1A. The theory of decision under certainty. This is embodied in the theory of the consumer's choice and the producer's choice as classically conceived. The assumptions are that the decision-maker has (i) perfect information, (ii) consistent goals, and (iii) 'well-behaved' opportunity sets.[5] It is then possible to deduce unambiguously one course of action, which is *the* rational thing to do. Removing assumption (iii) means that there may be *several* alternatives that are equally and maximally good, or (a more disturbing case) that there may be *no* optimal course. Removing assumption (ii) makes it impossible to compare alternatives, and *a fortiori* to choose the best among them. Removing assumption (i) can be done in three distinct ways, which are discussed under (1B), (1C), (1D) below. A simple example of decision under certainty is the engineer calculating the shortest trajectory for a rocket to the moon.

1B. The theory of decision under risk. Here the information is assumed to be imperfect, but still quantifiable, in the sense that for each possible course of action there is a known probability distribution over the set of outcomes. A simple example is the farmer choosing a crop mix for next year's harvest taking account of the known probabilities for each type of weather and the known properties of each crop in each type of weather. This example also serves to bring out a fallacy that must be avoided when discussing rational criteria for decision-making under risk. If the farmer deals with cash crops only, it might be tempting to advise him to choose the crop mix with the largest expected market value. This, however, might be bad advice because the farmer might be subject to a *risk aversion* that makes it rational for him to take account of the dispersion around the average (or expected) value and not only of this value itself (Roumasset 1976). See also Section XIII below.

1C. The theory of decision under uncertainty. This is a more radical conceptualization of ignorance. The actor is assumed to know the possible outcomes of any given course of action, but unable to assign any probability – or any range of probabilities – to these outcomes. This at once raises two related questions, which are discussed in Section X below. (i) How are we to distinguish the abstractly possible outcomes from the 'really possible' without invoking some notion of quantitative probability? (ii) Why should we not exploit the knowledge used for defining the set of outcomes, to assess their probability? The criteria for rational choice under uncertainty are rather complex in the general case,[6] but quite simple in the special case where all alternatives have the same 'best-consequence'. In that case we only have to compare the 'worst-consequences' and choose the alternative having the *best worst-consequence* (the maximin criterion). I argue below that this case obtains in the energy problem.

1D. Game theory (the theory of interdependent decisions). In the preceding cases the environment of the actor is supposed to be knowable (even if possibly unknown) independently of the decision to be taken. The environmental variables 'have' some values, regardless of the state of our knowledge about them. They are *parameters* for the decision-problem, whence the term 'parametric rationality' for such cases. If the environment is itself made up of other actors seeking to decide rationally, this no longer holds. For the individual actor it would then be irrational to look at himself as a variable, and at everyone else as parameters. Rather he must look at the problem as one of interdependent decision-making where all values of the variables are determined simultaneously. This 'strategic rationality' is explored by *game theory*. The basic structure of the theory can be summed up by saying it is a conceptual scheme that permits us to understand social interactions where (i) the reward of each depends upon the reward of all, (ii) the reward of each depends upon the choice of all, and (iii) the choice of each depends upon the choice of all.[7] As there is little scope for applied game theory in the energy problem,[8] I shall not here spell out the rather complex decision criteria that have been proposed for strategically rational actors.[9]

2. Collective choice theory. In this branch of decision theory the perspective is radically different. The problem is the following: given a set of individual rankings of a set of alternatives, how does one arrive at the social ranking? In traditional political philosophy one has imposed certain substantive demands on the process of aggregating the individual rankings as well as on the social ranking itself. To be precise, it has been required (i) that the aggregation procedure should be *democratic* and (ii) that the outcome of the process should be *just*. In addition one has sometimes required (iii) that individual preferences be *rational,* in the sense of not being swayed by narrow egoistic motivations. In collective choice theory only condition (i) is retained, and in addition one has required (iv) that the aggregation procedure should be rational, in the sense of being internally consistent and efficient, and (v) that the outcome should also be rational, in the sense of internal consistency.[10] The main result of the last 25 years' work can be summed up by saying that the requirements of a rational process, a rational result, and a democratic process are more than sufficient to determine the social ranking, so that no room is left for considerations of justice. *More* than sufficient, in the sense that the three sets of conditions, suitably specified, are incapable of simultaneous satisfaction (Kelly 1978). One way out has been to reintroduce some variety of condition (iii) above, though much more needs to be done in this direction.[11] Another way out has been to weaken the condition of democracy, or more precisely to drop the condition 'one man, one vote'. This can be done in several ways, each of which corresponds to a particular conception of social justice.[12] Collective choice theory can be fruitfully applied to the nuclear issue,[13] but I shall not here explore this potential.

III. The nature of the energy choice

On the assumption (discussed in the next section) that Western societies will need increasing amounts of energy, the main ways of satisfying future needs are fission power, fossil power, and hydroelectric power, singly or in combination. The doubts concerning sun, wind and water are too great for these to be more than interesting side options. Fusion power, which for a long time was thought to be the energy form of the future, has turned out to be a slow developer and will probably not be much cheaper than the alternatives (Holdren 1978, Parkins 1978). The restriction to fission, fossils, and hydroelectricity seems, therefore, to be justified. The problem then is: which combination of the three energy forms should be chosen? I shall refer to any such combimation as an energy *vector*.

In some discussions it is assumed that the decision process should be a sequential one: first we decide how much energy should be produced, and then we choose a suitable vector. This is clearly false. The two choices must be made simultaneously, because some forms of energy production may be so expensive or dangerous that we should avoid a total energy consumption so large that they will be required.

The choice of an energy vector can be made within more or less narrow assumptions about 'everything else'. The most restrictive method is to freeze the social structure, including income distribution, consumption patterns (housing, cars), centralization etc. Within this given framework we may then go on to ask which energy vector gives the best result, cost-benefit-wise. This gives the vector as a *partial maximum*. A somewhat broader method is to permit small variations and adjustments in the variables that were kept constant in the first procedure, so as to obtain the vector as a *local maximum*. The most general method is to permit large and simultaneous variation in all variables, so as to achieve the *global maximum*.

Using techniques such as 'piecemeal social engineering' (Popper 1957) or 'incremental planning' (see van Gunsteren 1976, Ch. 2), one has no guarantee that the maxima that are attained have a global character, nor that they are anywhere near the global maximum. The myopic policy whereby at any given time one climbs a gradient towards the nearest maximum may be disastrous in the long run (Elster 1979, Ch. I). Still there are several strong reasons for not experimenting with alternatives that are too distant from the present state. First, the imperfect knowledge that – under conditions of risk aversion – makes for a cautious approach (Section XIII). Second, the irreversibilities that make for caution even in the absence of risk aversion (Section XIV). Third, the endogenous change of preferences (von Weizsäcker 1971, Elster 1979, Ch. II.6) that should make us distrust our evaluation of future states which we have never experienced. Fourth, the problems of the transition period: both the inherent ills of social unrest and dislocation, and the political instability that might follow. Drastic proposals may be politically impossible (Elster 1978a, Ch. 3). To be sure, one should be very cautious in introducing political constraints in a model that is to serve as a basis for political decision.[14] If the analyst is to be of any use to the politicians, he should not endogenize their decisions (but see Lindbeck 1976). In the next section, therefore, I briefly discuss some proposals that appear to be politically impossible – at least until someone breaks out of the charmed circle of routine politics and redraws the map of the possible (Elster 1978a, p. 50–1).

Two final remarks concern the importance of *space* and *time* in structuring the energy choice process. Three spatial levels are involved: the local, the national, and the international. Conflicts of interests are possible between the first and the second and between the second and the third. To the extent that there is a trend towards community veto rights on power projects, we get the 'populist dilemma' discussed in note 13. A more obvious difficulty is the free-rider problem that arises if each local community wants to have the benefits from energy production (i.e. wants the energy) without the inconveniences that follow from accepting the plant on its territory. If each community expects someone else to take the burden, too little energy will be produced relatively to the national need. On the other hand, the national need may imply an excessive production relatively to the needs of the international community. The individual nation will be but little harmed by *its own* contribution to the pollution and harmful transformation of the atmosphere, even if the collective result of such behaviour on the part of all nations is collectively disastrous. It would be nice if these two conflicts annulled each other, so that community egoism could substitute for international solidarity. (Elster 1978, pp. 129–30, has a discussion of a similar case of two Prisoner's Dilemma-situations cancelling each other in their effects.) This, however, will only be the case if it is the *same* energy forms that produce the local and the international inconveniences, and there is no reason to think this to be generally true.

The temporal dimension is what distinguishes the energy choice from most or all other political choices. I can think of no other (civilian) problem where the relevant temporal horizon is so extremely large. Nuclear wastes will remain dangerous for a very long period. The plutonium in nuclear plants will forever remain accessible for the construction of nuclear weapons. Genetic defects may – because modern societies have deliberately broken the link between reproductive adaptation and ecological adaptation (Elster 1979, Ch. I) – perpetuate themselves to the end of time.[15] The paramount importance of uncertainty in the energy choice is due to this temporal dimension. The further into the future we apply our theories, the less they provide of numerical estimates.

IV. The need for energy

I will conduct the discussion on the realistic assumption that there will be a substantial increase in the need for energy. It should be mentioned, however, that this assumption may be questioned on two grounds. First, it has been argued that it is possible to have economic growth without growth in energy consumption. There is general agreement that this holds in the short run, and some authors have argued that it is valid also in the intermediate run for which present decisions are taken. An American survey by the Demand and Conservation Panel of the Committee on Nuclear and Alternative Energy Systems (1978) concludes that it is possible to double GNP from today to 2010 without any growth in energy consumption. It should be observed, however, that for nations that do not get most of their energy from hydroelectric plants this result is less encouraging than might be thought. For such nations zero growth of energy consumption does not imply that they can dispense with new energy sources as the old dry up. The problem might arise, for example, whether to go on using increasingly costly fossil fuel or to switch to nuclear power.

Second, and more radically, it has been argued that we should switch to a zero growth economy. Even in this case, of course, the remarks at the end of the preceding paragraph would apply, but the problem would be less serious if one could settle for zero or even negative growth of energy consumption. This conclusion is reinforced by the observation that zero growth in the production of goods and services may go together with a positive growth in welfare levels. Just as it is possible to dissociate energy growth from GNP growth, it is possible to dissociate GNP growth from welfare growth. The latter separation could be brought about by the abolition of the *positional goods* (Hirsch 1976) that are so important in modern economies. If everyone is motivated by the desire to be ahead of the others, then everybody will have to run as fast as they can in order to remain at the same place. *Without any change in preferences*, welfare levels could be raised if everyone agreed to abstain from this course (Haavelmo 1970). By contrast, most proposals to distinguish between 'real' and 'false', or 'natural' and 'social needs', imply that preferences should be changed – which immediately raises the spectre of paternalism. One should not confuse the needs that are social in their *object* (positional goods) with needs that are social in their *origin* (Cohen 1978, p. 103).

It is not inconceivable that an imaginative politician might find a way to abolish the race for positional goods, but in the meantime it is important to discuss what should be done before abolition. The inevitable suggestion that such discussions will reduce the incentive for abolition has some substance to it, but cannot be decisive. I return to this problem in Section X below.

V. Forecasting technological change

Any choice can be conceptualized as two successive filter processes. The first is the definition of the feasible set, i.e. the set of all the alternatives that satisfy the structural constraints. Such constraints may be self-imposed (Elster 1979, Ch. I) or 'given' in some sense. The second filter then singles out the member of the feasible set that is to be realized. The main emphasis in this paper is on rational choice as one such secondary filter; here I briefly discuss the first phase of the process.

Three constraints are often seen as given: climate, resources and technology. This is slightly misleading. Climate, in fact, interacts with energy consumption, being both a determinant of the need for heating energy and in part a consequence of energy use (through release of CO_2). Resources are given in some absolute sense, but not in the more interesting sense of available resources, availability depending upon extraction technology. And technology, in turn, obviously is not given, but subject to constant change. The real issue is whether we can predict or control the rate, nature and direction of technological change (Rosenberg 1976, David 1975, Nelson and Winter 1976).

I shall not here deal with the substantial problems of this issue, but only offer the following remarks. First, even if we accept that the details of change can neither be predicted nor controlled, it is possible to take account of the general fact that techniques *will* change. It may be unsound to tie down large amounts of capital when we know that new methods will become available. I return to this in Section XIV. Second, if we believe prediction and control to be possible, we should hold this belief in a consistent manner. One should not, for example, be an agnosticist in the question of developing safer methods for waste disposal and an optimist in the question of developing low-energy technology; nor, of course, should one have the opposite set of attitudes. Third, the first argument takes some of the bite out of the second, at least as regards the opponents of nuclear power. There is an asymmetry between proponents and opponents, because it will remain possible to change one's mind and opt for nuclear power if the soft technologies are not developed, whereas the adoption of nuclear power may to a higher extent be an irreversible one.

VI. The risk model for the energy choice

These preliminary hurdles being cleared, I now come to the substance of the problem. Let me first state that we can compare energy vectors (actually I shall only compare the pure energy forms) by looking at the inconveniences that are attached to them, because the economic benefits are of *roughly* the same order. If accidents never happened and there were no environmental effects, nor any problems of terrorism, sabotage or weapons proliferation, then fission power, fossil power, and hydroelectric power would have instalment and maintenance costs of roughly the same magnitude. (It is not essential for my argument that they be identical, only that the difference between the benefits be substantially smaller than the difference between the dangers and hazards.) Only with the advent of fusion power, as traditionally and not very realistically conceived, would there be an alternative with substantially smaller costs. A brief checklist of dangers and inconveniences of the various modes of energy production runs as follows:

– Hazards associated with the normal operation of power plants. For nuclear power this means release of radioactive material and possibly also of noble gases into the atmosphere. For fossil plants this means release of SO_2 and CO_2 into the atmosphere; also airborne radiation to an extent that may be greater than that from a nuclear plant (McBride et al. 1978).
– Accidents. For nuclear power plants this means reactor accidents following core meltdowns. For hydroelectricity it means dam accidents,

which occur with surprisingly high frequency (Mark and Stuart-Alexander 1977).
- Destruction of the natural landscape, especially with hydroelectric constructions.
- The dangers associated with nuclear waste disposal.
- The danger of sabotage of nuclear plants or dams.
- The danger of theft of plutonium for the purpose of making a nuclear weapon.
- The danger of proliferation of nuclear weapons when new nations get nuclear power plants, the plutonium for which can be used for weapons construction. Even if these nations could make bombs outside their nuclear power programmes, the latter 'can directly contribute to a process of "latent proliferation" whereby nations move inexorably closer to a weapons capability without having to declare or decide in advance their actual intentions' (Feiveson et al. 1979, p. 330).

In order to evaluate and compare these dangers many writers (e.g. the Rasmussen report) use a *simple risk model,* with the following structure. We assume that we face the choice between energy forms $E_1, E_2 \ldots E_n$, which all produce energy at the same cost if the above dangers are disregarded. We further assume that energy form E_i involves the dangers $F_{i1}, F_{i2} \ldots F_{im}$, and that the probability that danger F_{ij} materializes is p_{ij}. Finally we assume that F_{ij}, should it occur, has negative consequences evaluated as S_{ij}. We then define the risk of danger F_{ij} as $R_{ij} = p_{ij} \cdot S_{ij}$, and the total risk of E_i as $R_i = R_{i1} + R_{i2} + \ldots R_{im}$. We then choose the energy form with the lowest risk.

Stanford Research Institute (1976) uses a *sophisticated model* that takes account of the fact that the probabilities p_{ij} are not given, but rather functions of the amount of money b used for prevention and control: $p_{ij} = g_{ij}(b)$. In continuation of this argument we may also assume that the harm S_{ij} done if F_{ij} materializes is a function of the amount of money c used for protection: $S_{ij} = h_{ij}(c)$. For given b and c the total costs then become ($b + c + g_{ij}(b) \cdot h_{ij}(c)$). Choosing b and c so as to minimize this expression, we get the real risk R_{ij}. This method is clearly superior to the simple risk model, assuming, of course, that the costs of prevention, control and protection are quite broadly defined. If one wants to establish an efficient control system in order to protect against sabotage and theft, then the costs should include not only wages for guards etc., but also the undesirable effects of living and working in a *controlled environment.* If, for example, workers in nuclear power plants are subject to harassing methods of supervision, the compensation for this should be added to their wages (if it is not already reflected in them); similarly for people living in the neighbourhood of such plants, at least if they lived there before the construction of the plant. Not only may stress be produced by the omnipresence of guards etc., but perhaps even more by the need for having at hand the various means that can reduce the risk once an accident has occurred. It is not irrational to feel anxiety because of the need to protect yourself against an event that will almost certainly not occur.

Even the sophisticated model, however, is far too simple. It overlooks three important elements in the situation; uncertainty, risk aversion, and irreversibility. The remaining part of the paper explores these elements. Because the basic assumption of the risk models is that it is always possible to arrive at a numerical estimate for the probability of occurrence of the dangerous events in question, and that it is rational to use these probabilities as the basis for risk evaluation, I begin with a discussion of probability. I here distinguish between objective, theoretical, and subjective probability. This terminology is to some extent misleading. In the basic sense I agree with the de Finetti-Savage school of probability that all probabilities are subjective, meaning that they relate to our knowledge about the world and not to the world itself. (In the present context we can safely disregard the 'objectively random' events of quantum mechanics.) Nevertheless it is possible to distinguish between different kinds of *sources* for probabilistic knowledge: objective frequencies, theoretical calculations, and subjective calibrations. My question, then, is whether these sources are all equally reliable.

VII. Objective probabilities

Objective probabilities derive from the consideration of the observed frequency of certain events in the past. If 1000 tosses of a coin have shown heads 603 times and tails 397 times, then we may with some confidence predict a 60 % chance for heads the next time around. The 'fault-tree' analyses of the probability for reactor accidents use objective probabilities of this kind. For each part in the design there is a known probability of breakdown, based on frequency analysis. On the assumption that these probabilities are independent of each other, it is then possible to calculate the probability of simultaneous breakdown in so many parts that a reactor accident ensues. Conversely one may plan the design with so much redundancy that the probability of the simultaneous breakdowns can be made as small as desired.

If the assumption of independence does not hold, we have the 'common mode' failures that have posed serious problems for the investigators of meltdown probabilities. An important case concerns human error. The probability that a technician forgets to pull switch A is not independent of the probability that he forgets to pull switch B. In addition to the non-independence we might also expect to find aftereffects, in the sense that the commission of one error changes the probability that the person concerned will make further errors (Feller 1968, pp. 119 ff.). These two problems make it rather unlikely that frequency analyses of reactor accidents can ever achieve a very high accuracy. This, however, does not mean that we are in the realm of uncertainty. As explained in Section XI below, situations exhibiting both risk and uncertainty can be reduced to risk if the two interact multiplicatively rather than additively. This is probably the case in the reactor problem.

VIII. Theoretical probabilities

Theoretical probabilities are derived from a scientific theory. In order to predict the weather tomorrow (February 22) we may look at the weather of February 22 in a number of preceding years, and make our predictions on a frequency basis. There is, however, a better method available: look at the weather of February 21, 20 etc. of this year, and use these data as inputs in a scientific theory that predicts the weather with greater accuracy. (This example, incidentally, shows that the distinction between theoretical and objective probabilities is not a hard-and-fast one, for even the objective method requires a theory of planetary motion that tells us to take the average of all February 22's rather than of all days *tout court.*) A meteorological theory must, of course, be confirmed by frequency analysis of the observations, but it is still superior to mere extrapolation of past observations.

Theoretical probabilities are important in many aspects of the energy choice. To assess the probability of cancer induction by plutonium (Hohenemser et al. 1977, pp. 31–2) or of temperature changes in the atmosphere through continued release of CO_2 (Siegenthaler and Oeschger 1978), a large body of controversial theory must be invoked. *Controversial* is the key word in the present context. If there are several available theories competing in the field, which one or ones should be used for risk evaluation? Let us first assume that it is somehow possible to assign probabilities for the truth of theories, so that theory A has a 70 % chance of being the correct theory, theory B a 25 % chance, and theory C a 5 % chance. A first reaction might then be that we should go by theory A and use the probabilities derived from that theory for risk assessment. On reflection this is clearly inadequate. If theory B or even theory C predicts that if we take a certain course of action, it is very probable that something disastrous will happen, whereas theory A assigns a negligible probability to this event, then it would be irrational not to take some account of the less probable theory. It is true that for pure research the scientist should put all his money on one horse, be it the most probable theory or the theory with the highest present rate of problem-solving (Laudan 1977), but for applied research (and when financing pure research) one should spread one's bets more evenly.

If theories are to be assessed probabilistically (as discussed in Hesse 1974), it must be done in some kind of Bayesian framework. Here we first assign prior probabilities to the theory, calculate the probability of the observations, given the theory, and then use Bayes' theorem to find the probability of the theory, given the observations. The weak link here is the prior probability attached to the theory. To fix this by the principle of insufficient reason is clearly inadequate, for, in the first place, this gives different results according to how we individuate theories and, in the second place, it is incompatible with the fact that we do have a great deal of knowledge – an excess of reasons, in fact – that is relevant for the assessment of theories. A theory should (i) fit in with our general metaphysical views, such as the principles of determinism, local causality, continuity etc.; (ii) be compatible with theories in adjacent or analogous disciplines; and (iii) have the properties of simplicity and consistency. The problem is how to integrate all this knowledge and all these requirements into an overall evaluation of the theory. Here some kind of subjective probability must be used, a notion that is further discussed in the next section. To anticipate, the subjective probabilities are often unreliable and should not be used as a basis for action. In the special case of attaching probabilities to theories, I believe that *if* the subjective method is to be used, one should apply it directly to the posterior probabilities and not use the roundabout method prescribed by the Bayesians. For the working scientist to whom one would have to appeal, attachment of probabilities to theories prior to observation would be a very strange procedure. The only way to tap his scientific judgement would be to include the data among the material used for the assessment. (Of course the probabilities thus elicited might be used as prior probabilities when new data emerge.)

There are, unfortunately, enormous problems in assessing the reliability of such probability assessments of theories. The better a scientist knows the field in question, the more he is likely to identify himself with one of the competing theories. This identification is both a condition for fruitful work – and a source of bias in the assessment. The impartial *and informed* specator does not exist in science.[16] For this reason the best procedure would often seem to be to renounce a ranking of the theories, or at least of the theories that pass certain operational tests for minimal plausibility. This means that in our choice between theories we are in a situation of uncertainty, even if each of the theories places us in a situation of risk. As far as I know this problem has not been raised in the literature, and it is hard to say what is the rational procedure in such cases.[17]

Actually the situation may be even more complex. It is suggested by Dagfinn Føllesdal (1979) that the larger the number of competing theories, the larger the probability that they are all false. If we have several theories that use rather different kinds of assumptions, this may be seen as a sign that the object of these theories is a part of the universe that we simply do not understand very well. It may then be rational to assume that something even worse may happen than what is predicted by any of the theories, or that the probability that the worse will happen is even larger than what any of them implies.

IX. Subjective probabilities

Many of the problems that are relevant for the energy choice do not lend themselves to an analysis in terms of objective or theoretical probabilities. The probability of attempted theft or sabotage is one such problem; the probability that a government at some time in the future will use plutonium from its nuclear plants to construct atomic weapons is another. Nevertheless there have been attempts (Stanford Research Institute 1976) to quantify these probabilities, presumably on the basis of subjective probabilities. The rationale behind the use of subjective probabilities – consistent hunches – is the following. It is probably never the case that we are in a state of complete ignorance with respect to a given problem. If we are able to state the problem, we must know something about it. Even for the most esoteric problems (is there life on other planets?) we have a large body of information that is somehow relevant. Even if this body has not been incorporated into a theory, it is there, and it would not be rational to act as if it were not available. What could be more irrational than to act as if we are ignorant when in fact we are not?

The argument is persuasive, but fallacious. It presupposes that we are somehow able to integrate our various fragments of knowledge and arrive at a stable and intersubjectively valid conclusion. The role of a scientific *theory* is precisely to effect such an integration, but in the absence of a theory all we can invoke is *judgement* – an elusive quality indeed. We can to some extent count on finding this quality in persons who would not have survived without it, such as successful politicians or businessmen (Elster 1979, Ch. III.4), but there is no reason to expect it in scientists or bureaucrats, on whom it would presumably fall to quantify the probabilities in question. There are too many well-known mechanisms that distort our judgment, from wishful thinking to rigid cognitive structures, for us to be able to attach much weight to the numerical magnitudes that can be elicited by the standard method of asking subjects to choose between hypothetical options. (See Tversky and Kahneman 1974, Janis and Mann 1977, and Slovic et al. 1977.) We should resist, that is, the implied step from 'is' to 'ought'. From the fact that it is always possible to elicit these subjective probabilities, we should not conclude that one ought rationally to act upon them. One could certainly elicit from a political scientist the subjective probability that he attaches to the prediction that Norway in the year 3000 will be a democracy rather than a dictatorship, but would anyone even contemplate *acting* on the basis of this numerical magnitude?

X. The uncertainty model for the energy choice

To the extent that the energy choice raises issues that cannot be analysed in terms of objective or theoretical probabilities, we are in a state of uncertainty, unless we are in the happy situation of being able to rely upon the subjective judgement of persons of proven reliability. And to the extent that we are in a state of uncertainty, it is rational to act as if the worst that can happen, is bound to happen. (Recall that this simple criterion is only valid when all alternatives have the same best-consequences, so that only the worst is relevant.) What does this imply in the context of the energy choice? Among the dangers listed in Section V above, uncertainties are important in many, indeed in most. As argued in the next section, however, uncertainty is swamped by risk in many of these problems. Genuine uncertainties remain in two cases: the proliferation of nuclear weapons and the heating of the atmosphere through release of CO_2 from fossil plants. About the first problem we have no theories, about the second too many.

At this point it is necessary to distinguish between different kinds of decision-makers. If the energy choice arises in a nation that already has acquired nuclear weapons, the nuclear weapons proliferation argument has no weight. And if the decisions are to be taken by the nation-state on the basis of the expected consequences *for that state only,* the CO_2 issue is irrelevant, as no nation – or only a very large nation – can be substantially harmed by the consequences for the atmosphere of its own fossil plants. For the CO_2 issue to be relevant, the nation-state must *either* act in terms of moral considerations such as the categorial imperative, *or* be so large that it can feel the impact of its own actions, *or* leave the decision to some kind of international body.

On the basis of pure self-interest, reinforced by moral reasoning, a nation that has not already acquired nuclear weapons should abstain from nuclear power, so as not to hand down to later generations a potential for starting or precipitating a nuclear war. As no one can say with any confidence whatsoever what the chances are that, say, Norway a thousand years from now will have a government that could be tempted by this potential, one should act as if it is certain that the potential would be exploited, and this implies that it should not be created. On the basis of moral reasoning (for large nations reinforced by self-interest) a nation should not extend its use of fossil power. For countries with a third option, extension of hydroelectric power, these are acceptable conclusions. For the majority of nations that cannot take this way out, there is a choice between (at best) zero growth, fission power, and fossil power.

The first has been ruled out by assumption, the second by self-interest, the third by morality. Something has got to give. In Section XV the empirical problem is raised: what is *likely* to be sacrificed: growth, self-interest or morality? To the extent that one can say something about what *should* be sacrificed, I submit that fossil power is less harmful than nuclear power. If there is a gradual increase of heat in the atmosphere, then at some future date there will be a political pressure that will make the first option – zero growth – politically possible or even politically necessary, even though this is not the case to day. It is politically possible to opt for a course that will make the best course politically possible at some later date (Elster 1978a, p. 52). With nuclear power, on the other hand, such a turnabout is impossible: if a nuclear war ever starts, it is too late to do anything. I freely admit that there is something strange in my advocacy of fossil power, and that my argument really leads up to the no-growth conclusion. We are dealing, however, with second-best arguments here, as is often the case in politics. In this particular case the second-best proposal can be justified by first-best considerations, as just explained. This should go *some way* towards making the argument more palatable to the groups who would otherwise characterize it as 'selling out to the establishment'.

Before leaving the uncertainty issue, I must answer a question that was raised in Section II above. If an event is uncertain, in the technical sense of the term used here, how do we know that it is really possible? We do not want to include on the list of consequences all kinds of abstractly possible events, such as invasion from outer space as a result of radioactivity emitted from nuclear plants and detected at Alpha Centaurus, but how can we distinguish between the abstractly possible and the really possible? I submit that one criterion is the following. As stated above, uncertainty does not imply total ignorance. On the contrary, we often have vast amounts of information that is somehow relevant to the event in question, the difficulty being one of evaluating the *net effect* of all this knowledge. In economic theory before the advent of mathematical models this was very often the case. One economist could say with confidence that tariff barriers would lead to increased employment, another that they would reduce employment, and they could both be right in the sense of being able to justify their assertions by a partial theory, describing a partial mechanism in the real world. On the other hand, they were unable to reach agreement, as they did not have at their disposal the mathematical tools that make it possible to determine the net result of the opposing mechanisms. It is well known from systems analysis (Forrester 1971, pp. 109–10) that such long-term net effects often are counterintuitive, confirming what I said above concerning the objections to subjective probabilities in such cases. Still the partial theories do have one use: they can circumscribe the really possible from the abstractly possible. Any consequence that is predicted by a partial theory that is valid *ceteris paribus* is a 'really possible' consequence. In order to say how probable that consequence is, we would also need to have a general theory of what happens when the *cetera* are not *paria*.

XI. Mixing risk and uncertainty

I have alluded to the problems that arise when a situation has elements of both risk and uncertainty. One of these problems has been left unsolved: how to act when we are uncertain as to the correct risk-analysis of the situation? This is a difficult problem because uncertainty and risk here are hierarchically arranged. The problems are more easily tractable when uncertainty and risk coexist at the same level. If for simplicity we limit ourselves to additive and multiplicative interactions between elements in the problem situation, then uncertainty dominates risk in additive interactions, whereas it is the other way around in multiplicative interactions. Let me explain these statements through some examples.

In additive interactions uncertainty dominates risk. Nuclear energy has the risk of accident and the uncertain consequence of nuclear weapons proliferation. It does not matter how small we make the former, as long as the latter is present. A large number + a small number remains a large number. However, a large number multiplied by a small number may be a small number, especially if we can make the small multiplicator as small as we want. Take the case of successful use of stolen plutonium to make atomic weapons, as analysed by Stanford Research Institute (1976). The probability of this event is the product of three probabilities: (the probability that theft will be attempted) × (the probability that the theft will be successful if attempted) × (the probability that the construction of the bomb will succeed, given the successful attempt). Here the first and third events are clouded in uncertainty, but the second has a quantifiable probability that will depend upon the security measures that are taken. (This statement abstracts, perhaps without justification, from the possibility that the thieves may have intra-mural helpers that could make these measures as irrelevant as was the Maginot line.) This means that the probability of the event as a whole is not greater than the probability of the second event, which can be assessed and even controlled.

The same situation seems to obtain in the nuclear waste problem:

Because we are unable to prove that there is no risk of water intrusion in these formations, we shall make the conservative assumption that groundwater flow will eventually take place. We will determine whether there are controlling parameters that may still ensure confinement and whether there are any consequences for the environment. We believe that it is more important to make these kinds of estimation when studying a repository than to try to assess probability coefficients for the occurrence of groundwater. (de Marsily et al. 1977, p. 521; see also Kubo and Rose 1973, p. 1207.)

XII. Acting as if the worst will happen

The maximin criterion for decision under uncertainty, to be used when all alternatives have the same best-consequences, tells us to compare the worst-consequences and then choose the alternative with the best worst-consequences. I have argued that nuclear power has the worst worst-consequence: a global nuclear war; that fossil power has a somewhat better worst-consequence: the reversible (or at least arrestable) heating of the atmosphere; whereas the best worst-consequence belongs to an option that at present looks politically impossible, the no-growth economy.

To compare worst-consequences in this way can be paraphrased as acting as if the worst will happen. This principle can also be defended in cases of risk rather than uncertainty, viz. if some of the consequences are infinitely worse than the others. If the extinction of mankind is evaluated at minus infinity, then it swamps all disasters of finite size. We must, of course, have some precise probability attached to this event: mere logical possibility is not sufficient. It does not matter, however, if this probability is extremely small, for an infinite number multiplied by any positive number remains infinite. This is Pascal's wager turned on its head. The weakness of this argument is that it may turn out that most actions have such total disasters associated with them, as an extremely unlikely but still quantifiable probability. I believe, therefore, that one should be very cautious in arguing along these lines. The uncertainty argument for the principle of acting as if the worst will happen is much more solid.

Some writers (e.g. the Swedish Energy Commission 1978) propose the following criterion: act so as to avoid a too great chance of disaster. To be precise, they suggest that if there is a probability $>\alpha$ of a disaster $>\beta$, then this course of action should not be taken. This is sometimes characterized as a variety of the maximin criterion, but clearly it is not. It is an operational version of the expected-utility criterion that may have something to be said for it in concrete applications, but that is too pragmatic to bear a great theoretical weight.

XIII. Risk aversion

The preceding section concludes my case for the uncertainty model, as opposed to the risk model. There are, however, two additional problems with the simple and sophisticated risk models of Section VI that are independent of the uncertainty issue. The first concerns risk aversion, the fact that we cannot automatically equate an estimated danger S_1 occurring with probability p_1 and an estimated danger S_2 with probability p_2, even if $p_1 S_1 = p_2 S_2$. To put it more concretely, one chance in a million of a million deaths and one chance in a thousand of a thousand deaths are not necessarily equivalent. The theory of risk aversion for individuals is highly developed,[18] but not necessarily relevant for social choices. Let me briefly point to one reason why society should prefer a large probability of a small accident to a small probability of a large accident, and one reason for having the opposite preference, assuming that the risks as defined above are identical.

The death of an individual is a tragedy for that person, for his friends and relatives, and for society as a whole. The first element cancels out in the present comparison, but the last two point in opposite directions. The death of a person is a tragedy for society because, or to the extent that, he is the bearer of specific skills or specific traditions. It is not very likely that a single individual will be irreplaceable, but quite likely that the destruction of a whole community implies the loss of a part of the national heritage. This, therefore, is one reason for preferring the large probability of a small accident. On the other hand, the number of relatives and friends that are bereaved will be smaller in a large accident, for here the odds are that many of these individuals will perish at the same time. This provides a reason for the opposite preference. I believe that the first argument – giving preference to large probabilities for small accidents – dominates when the small accidents become relatively large, for then the second argument will not distinguish very sharply between the two cases. One may prefer, that is, a chance in a thousand for the death of a thousand persons both to certain death of one individual and to a chance in a million for the death of a million persons. This is grisly reasoning, but then these are grisly matters.

XIV. Irreversibility

Claude Henry (1974) has shown that even with risk-neutrality, the risk models may not be adequate for the analysis. Expected value may not be the best maximand when irreversibilities are present, as they often are in the energy choice. The reason for this was briefly alluded to in Section V above, and will now be spelled out more fully. Let us assume that we have a choice between two alternatives: to make an irreversible decision and to postpone the decision. We assume that given the present information, the first alternative has the largest expected value. On the other hand we do have information (of a quite general kind) that we will get more information (of a specific kind) that is relevant to the decision. (We do not know, of course, what information we shall get, for if we knew that we would already have it.) If at this later date we get even better reasons for making the decision, we will not have lost too much by postponing it, for a decision of postponement can usually be reversed. If, on the other hand, the later information provides arguments against the decision, then it would be better if we had not taken it. This argument, of course, must not be stretched too far, for then irreversible decisions would never be taken. The argument only makes sense within some theoretical framework that specifies the structure of expected information. Nevertheless it is intuitively obvious that the point is important. (Observe that it is also related to the problem discussed in the last paragraph of Section VIII.)

In the energy debate the irreversibility problem has been raised especially in connection with nuclear waste disposal. Here irreversible techniques have partly been seen as a solution to the problem of an irresponsible future, and partly as themselves raising the problem of premature commitment discussed above (Kubo and Rose 1973, Rochlin 1977). Environmental effects of CO_2 have also been discussed under this heading. Let us, however, distinguish between two forms of irreversibility. For this purpose it is useful to have the notion of a *frontier*, the value of some variable above which there are disastrous environmental or social effects. *Strong irreversibility* then obtains if (i) one can only know where the frontier is by hitting it, and (ii) it is impossible to back away from it when you hit it. *Weak irreversibility* obtains when condition (ii) is satisfied, but not condition (i). The first condition is a very strong uncertainty condition, saying that not only are we ignorant now about how far we can go, but that we will remain so until we have gone too far. 'You never know what is enough unless you know what is more than enough' (Blake, 'Proverbs of Hell'). It may be the case that CO_2 effects are weakly irreversible (Siegenthaler and Oeschger 1978, p. 394), but they do not seem to be strongly irreversible. By contrast a nuclear war is strongly irreversible. The fulfilment of condition (ii) is obvious. As for condition (i), history tells ut that brinkmanship and provocation differ mainly after the fact.

XV. Politics and decision theory

I conclude with some remarks about the relation between the normative and the empirical aspects of the energy choice. On this issue I can only offer two rather banal observations, based on casual evidence rather than systematic research. The first is that engineers and economists seem to think rather differently about these problems. The engineer tends to think in terms of risk and of quantified probabilities, whereas the economist by training is more open to the possibility of genuine uncertainty. I therefore propose this as a working hypothesis: the more engineers are prominent in the decision process, the more likely are risk models to be adopted and the more resistance there will be to arguments based on uncertainty. To the engineer, uncertainty is a challenge to get more knowledge. This in itself is a good attitude, but not if it blinds itself to the possibility that there are some issues where it is inherently impossible to get more information.

The second observation concerns the reason why politicians may not be much swayed by considerations of the kind advanced here. To demand of a politician that he accord large weight to the welfare of later generations and/or to the welfare of other countries, is rather like asking him to commit suicide. The most important issues in the nuclear power issue concern the distant – perhaps the very distant – future. The most important problem about fossil power relates to the impact on other countries of CO_2 release in the atmosphere. By contrast the environmental effect of dam construction for hydroelectric power is immediately visible here and now. In Norway we have the paradox that political pressures from the ecological and environmental groups, mediated by the youth organizations of all political parties, may lead us to abstain from extending the hydroelectric power system, even if this is vastly superior to the other alternatives according to the value systems of these very groups. This is so because direct environmental destruction tends to mobilize more people than actions that have consequences more remote in time or space.

Let me add, on a tentative note, that morality may prove stronger than long-term self-interest. Morality has *some* present advocates, viz. the system of international organizations, whereas the spokesmen for later generations are few and far between. To the extent that this is true, I would expect nuclear power to be chosen over fossil power, even if the opposite would seem a more rational choice. Nevertheless I believe that the fossil option is a genuine political possibility.

APPENDIX 2

The contradiction between the forces and relations of production

With a Mathematical Note by Aanund Hylland*

1. Introduction

In this Appendix I offer what I believe to be a new interpretation of Karl Marx's theory of the relation between productive forces and relations of production in capitalism. The analysis is intended to be exegetically faithful; indeed I believe that it makes it possible to reconcile texts that on the standard interpretation appear to contradict each other. I also hope that it may be of some interest in its own right. In particular, it provides a conceptual framework for the empirical and welfare-theoretical questions relating to the transition from capitalism to communism.

In Section 2 I quote some of the important texts that bear upon the problem, and explain intuitively the difficulty that confronts the interpreter. In Section 3 I set out the model that specifies the interpretation. Section 4 offers some comments on the model, and in particular on the problem of finding the optimal transition point from capitalism to communism. The Mathematical Note contains the proofs of some assertions made in Sections 3 and 4.

2. Marx on productive forces and relations of production

Few will deny that historical materialism is at the heart of Marxism, or that the relation between productive forces and relations of production is

* I am grateful to G. A. Cohen, Leif Johansen, and Gunnar Opeide for their comments on an earlier version of this article. I should like to stress that the contribution of Aanund Hylland goes much beyond the Mathematical Note. On the basis of my early outline of the axiom system, he showed how it could be simplified and how the theorems (for which he has full responsibility) could be derived. He also made valuable comments on the non-mathematical parts.

at the heart of historical materialism. One might, therefore, expect Marx to have explained in detail and with some precision what these notions mean and how they are related to each other. At the very least we might expect such elaborations to be forthcoming for the special case of the capitalist mode of production, the case by far the most intensively studied by Marx. As all readers of Marx know, such was not his wont. The meaning of the terms and of the theory relating them to each other must be reconstructed from scattered texts, written over some 30 years. The basic text for our problem is in the Preface to the *Critique of Political Economy:*

> In the social production of their life, men enter into definite relations which are indispensable and independent of their will, relations of production which correspond to a definite stage of development of their material productive forces. . . At a certain stage of their development, the material productive forces of society come into conflict with the existing relations of production, or – what is but a legal expression for the same thing – with the property relations within which they have been at work hitherto. From forms of development of the productive forces these relations turn into their fetters. (Marx 1859.)

Here there are four key terms to be interpreted: 'forces of production', 'relations of production', 'correspondence', and 'contradiction'. The passage asserts that within each mode of production two stages can be distinguished. In an initial stage there is correspondence between the productive forces and the relations of production, and in a later stage the relation is one of contradiction. Nothing is said concerning the mechanism that brings about the change from correspondence to contradiction. The exact nature of this mechanism in the case of capitalism is the topic of the present paper. Before we can enter that subject, however, we need some understanding of the terms involved.

For the fine grain of the analysis of 'productive forces' and 'relations of production', I refer the reader to Cohen (1978). I stipulate without further discussion that relations of production are defined in terms of the property of the means of production, including the labour power involved. Something more needs to be said, however, about the productive forces. I shall assume that these are essentially identical with the given technology, including the labour skills and the scientific knowledge associated with it. But Marx uses the terms *'Produktivkraft'* and *'Produktivkräfte'* ambiguously, sometimes stressing the abstract productivity embodied by the technology and sometimes the physically concrete embodiment. When referring to the *development* of the productive forces, Marx certainly refers to the abstract quantitative aspect. But technologies also undergo *change* and exhibit *differences* that can be described in purely qualitative terms. And it seems that Marx and his interpreters sometimes refer to the qualitative aspects of technology when discussing the correspondence-contradiction problem.

The twin terms 'correspondence' and 'contradiction' appear from *The German Ideology* onwards in basically the same contexts and therefore, presumably, in basically the same sense. This sense, however, proves very elusive. The central difficulty is the following. The 1859 text quoted above makes a series of general statements, intended to be valid for all modes of production, from the Asiatic one up to and including capitalism. On the other hand, these statements, in their most natural (or at least standard) interpretation, seem to be contradicted, in different ways, both by Marx's statements about pre-capitalist modes of production and by his account of capitalism. I shall argue that the first inconsistency is a real one. In Marx's analysis of pre-capitalist societies there seems to be no room for the change from correspondence to contradiction affirmed by the general theory. By contrast, I believe that the interpretation is capable of eliminating the second inconsistency.

Let me first set out what I have called the most natural or standard interpretation of the Preface. It is that in any mode of production there is initially a high rate of technical progress, which ultimately slows down to stagnation. The reader may visualize the process as the familiar logistic curve, and the change from correspondence to contradiction can then be identified as the point at which the rate of technical progress begins to fall. But these details are not really important. What is important, is initial growth and ultimate stagnation within each mode of production. The standard interpretation then affirms that the relations of production change when and because technical stagnation sets in, and that they are replaced by a new set of relations that once again permit technical progress. For examples of this interpretation, the reader is referred to Plamenatz (1954, p. 20, p. 28), Kolakowski (1978, vol. I, p. 375), and Cohen (1978, p. 173, proposition (4)). I believe that this interpretation is wrong, but it would be less than honest to fail to mention the textual basis for it in the Preface: 'No social formation ever perishes before all the productive forces for which there is room in it have developed.' I believe that this single sentence cannot have priority over the other texts, to be mentioned later, which affirm that capitalism *will* perish before all the productive forces for which there is room in it have developed.

An important feature of the standard interpretation is that it asserts that the notion of correspondence is a *dynamical* one, i.e. implying a technical development. It is in fact a logical requirement of the theory that the notion of correspondence shall have this dynamic character, at least if the contradiction is to emerge endogenously from the correspondence. And this emphasis on endogenous change is, I believe, central to historical materialism. To see the importance of this dynamic definition, consider two alternative – and not uncommon – definitions of the correspondence as a purely static notion. First, there is the temptation to see the contradiction between productive forces and relations of production as a suboptimal *use* of the former, with the implication that the correspondence consists in optimal use. But optimal use is a static notion that cannot in any way engender its opposite. If at some point in time the relations of production are such as to permit optimal use of the existing technology, then this has no implications beyond the given point in time. Secondly, there is the temptation to explain the correspondence in terms of the qualitative character of the productive forces, as when one argues that the need for irrigation can explain the relations of production independently of the level of productivity. It may be true that irrigation requires coordination and centralization of some economic decisions, but once again we are left in the dark as to how this correspondence could contain the seed of its own destruction. By contrast, if correspondence implies a change in technology, then we can perceive the possibility that the correspondence might thereby be destroyed.

The second crucial feature of the standard interpretation is that it asserts that the notion of contradiction is a static one, implying that change has come to a standstill. While I retain the first feature of the standard interpretation, I reject the second, as explained below.

The standard interpretation, then, asserts growth followed by stagnation within each mode of production. This is contradicted by Marx's statement that pre-capitalist modes of production exhibited uninterrupted stagnation; and also by his view that capitalism exhibits uninterrupted (and in fact accelerating) growth. The first contradiction should be blamed on Marx, and not on the standard interpretation. The second, however, can be eliminated by improving on the standard interpretation.

In the *Communist Manifesto* Marx writes that 'The bourgeoisie cannot exist without continually revolutionising the instruments of production, and thereby the relations of production and the social relations. Conservation, in an unaltered form, of the old modes of production was on the contrary the first condition of existence for all earlier industrial classes.' The passage is also cited in *Capital*, in a footnote to a sentence stating that 'all earlier modes of production were essentially conservative' (Marx 1867, p. 486). In fact, Marx believed that pre-capitalist modes of production required for their perpetuation an unchanged set of productive forces, and that they burst asunder when the productive forces developed beyond the small compass compatible with the existing relations of production. In other words, here correspondence means technical stagnation, and contradiction means technical change, a complete reversal of what the 1859 Preface asserts on the standard interpretation that I have just set out. The important text from the *Grundrisse* where Marx exposes his view on pre-capitalist modes of production should be quoted in full:

Although limited by its very nature, [capital] strives towards the universal development of the forces of production, and thus becomes the presupposition of a new mode of production, which is founded not on the development of the forces of production for the purpose of reproducing or at most expanding a given condition, but where the free, unobstructed, progressive and universal development of the forces of production is itself the presupposition of society and hence of its reproduction; where advance beyond the point of departure is the only presupposition. This tendency – which capital possesses, but which at the same time, since capital is a limited form of production, contradicts it and hence drives it towards dissolution – distinguishes capital from all earlier modes of production, and at the same time contains this element, that capital is posited as a mere point of transition. *All previous forms of society foundered on the development of wealth, or, what is the same, of the social forces of production.* (Marx 1857, p. 540.)

Marx here contrasts the development of the productive forces under capitalism with their development both in pre-capitalist society and in the higher communist society. I return to the second of these comparisons below. In the phrase that I have italicized (and in which I have corrected a fatal error in the Nicolaus translation) Marx says that pre-capitalist modes of production foundered because they were unable to absorb technical change, and he goes on to quote the destruction of feudalism through gunpowder and printing (among other things) as an example. I believe that it is impossible to make this consistent with the standard interpretation of the Preface. In particular, I cannot accept the recent proposal of Cohen (1978, pp. 169ff.) that the conservatism of earlier modes of production can be made consistent with the Preface by a distinction between two senses of the term 'forms of development'. Cohen says that the relations of production may be seen either as forms *by means of which* the productive forces are developed or as forms *within which* they develop. Pre-capitalist relations of production were not forms of development in the first sense, i.e. they offered no direct stimulus to the growth of the productive forces. But, Cohen argues, they were forms of development in the second sense, so that no other set of relations could have made the productive forces grow faster at that stage of their development. The distinction is explained by an analogy with constitutional monarchy, which may promote indirectly the development of democracy in a society, even though its direct effort is to oppose it. Now this argument suffers from two defects. First, it is inconsistent with what Cohen (1978, p. 292) says about the way in which productive forces 'select' relations of production, which is through the class that is best able to guarantee a stable and thriving production. This mechanism presupposes the first sense of the term 'forms of development'. Secondly, Cohen's argument is inconsistent with the text from the *Grundrisse* quoted above. Here Marx says that the pre-capitalist relations were destroyed when and because the growth of the productive forces was *too fast,* and this does not make sense on Cohen's view. On that view, the relations of production in pre-capitalist societies change when they are no longer optimal for the development of the productive forces, and this cannot be made consistent with the view that the relations change when and because the productive forces develop too much.

I now turn to the development of the productive forces under capitalism. Before entering the exegetical discussion, let me briefly observe how in this case there is a mechanism that could sustain the standard interpretation, of initial growth endogenously bringing about subsequent stagnation. Consider the following four-stage argument. (i) The conditions of perfect competition that prevailed under early capitalism encouraged a high rate of technical progress. (ii) Technical change being typically labour-saving, this development made for important economies of scale. (iii) Economies of scale lead to oligopolies and imperfect competition. (iv) Imperfect competition implies a lower rate of technical change. Of these, only (iii) can be accepted without qualification, as one cannot simultaneously have economies of scale, profit-maximizing entrepreneurs, and perfect competition. Statements (i) and (iv) are partially endorsed by Arrow (1971), but with reservations for the greater capacity of oligopolies for internalizing the effects from non-patentable innovations. Statement (ii) is, of course, very controversial. Nevertheless I believe that this is the kind of mechanism that one would have to look for if the standard interpretation were correct. It will not do to invoke the wasteful, inefficient or destructive *use* of the technology, since use and development are two quite distinct notions. Indeed, economists from Schumpeter to Arrow have argued that the dynamic optimality of capitalism in developing the productive forces depends in part on its static suboptimality in using them efficiently. The patent system is a well-known example. In other words, one cannot explain correspondence as optimal development and contradiction as suboptimal use of the productive forces.

But I believe that the standard interpretation is wrong, so that there is no need to search for mechanisms of the kind sketched above. There is not, to my knowledge, any passage in *Capital* or elsewhere in which Marx asserts that the rate of technical progress is declining under capitalism. Nor, it is true, do I know of any direct statements to the contrary. There are, however, statements that jointly imply that the rate of technical change is increasing. First, Marx asserts that in the course of capitalist development 'the rapidity of the change in the organic composition of capital, and in its technical form increases' (Marx 1867, p. 631). Secondly, 'the level of productivity attained is manifested in the relative preponderance of constant over variable capital' (Marx 1894, p. 759). Indeed, this is but one of the many passages in which Marx tells us that the increase in the productivity of labour can be measured by the increase in the organic composition of capital, as can be verified by looking at the index in Marx (1863), vol. III, p. 636, under 'Zusammensetzung des

Kapitals, wiederspiegelt den Stand der Arbeitsproduktivität'. Taken together, these statements imply an increasing rate of technical change under capitalism.

Other passages suggest the same conclusion. Thus in his discussion of the falling rate of profit, Marx (1894, p. 249) stresses the 'development of the productive forces of labour at the expense of already created productive forces'. (Indeed, models have been proposed in which the rate of profit falls because the capitalists consistently underestimate the rate of technical progress, and so are led to scrap new machinery ahead of the expected time. See Roemer (1981, pp. 123–4) for brief comments on these models.) More generally, unlike the many economists of his time who saw technical progress mainly as a counteracting force to the fall in the rate of profit, Marx saw it as the main mechanism by which the fall took place. The contradiction of capitalism is, for Marx, that technical change, while introduced to halt the fall in the rate of profit, only contributes to accelerating it (Elster 1978a, pp. 113ff.). True, in his discussion of the falling rate of profit Marx does not say that the rate of technical change is increasing, but the general tenor of his analysis definitely is incompatible with the idea of technical stagnation causing the downfall of capitalism. I return below to the relation between the 'contradiction' involved in the falling rate of profit and the 'contradiction' between productive forces and relations of production. The reference in this paragraph to the falling rate of profit was made only to back the exegetical contention that Marx did not think of technical change under capitalism as stagnating.

Recall now what Marx says in the passage from the *Grundrisse* quoted above, concerning the future society in which 'the free, unobstructed, progressive and universal development of the forces of production is itself the presupposition of society and hence of its reproduction'. Behind this statement, and many others in the same vein, lies the image of man worked out in the *Economic and Philosophic Manuscripts of 1844*. According to this view, innovative and creative activity is natural for man, and springs from the inner sources of his being. Contrary to the usual approach in political economy, the problem is not one of creating the incentives to innovate, but of removing the obstacles to the natural innovative drive of the individual 'in whom his own realisation exists as an inner necessity' (Marx 1844, p. 304). Special incentives are required only in conditions of scarcity and poverty, in which the needs of the individual are twisted and thwarted. In the early stages of capitalism there was indeed a great deal of scarcity and poverty, and unavoidably so, as the material conditions for a high level of want satisfaction were not yet created. Under these conditions, for example, the patent system was the best and most progressive arrangement, even though it subordinated progress to profits. This system, however, inevitably created the conditions for its own demise. In later phases of capitalism there is still a great deal of poverty, but avoidably so. Given the technology developed by capitalism itself, it is materially feasible to install a regime in which the level of want satisfaction is so high that innovation as a spontaneous activity comes into its own – at a rate far in excess of anything that has existed before.

This, I believe, is the essence of the 'contradiction between productive forces and the relations of production' under capitalism. The capitalist relations of production tend to make themselves superfluous, by bringing into being productive forces that require new relations for their further optimal development. This interpretation fits both the 1859 Preface and the statement of constant or even increasing rate of technical progress under capitalism. It is quite possible for technical progress during the later stage of capitalism to be *both increasing and lower than what it would have been under a socialist regime starting at the same technical level*. This idea is made precise in the model presented in the next section. Here I only ask the reader to go back to the 1859 text, which in all probability he has already studied to the point of not really seing what it states. He will then see that it does not at all assert that technical progress in the later stage is slower than it was during the early phase. The base-line for the comparison can equally well be a counterfactual one, viz. the rate of growth that would have obtained in a different institutional setting, given the same initial technical level.

3. The interpretation

I here assume that 'development of the forces of production' means technical progress, as measured by the rate of growth of productivity. The level of technology at time t during the actual or hypothetical development of capitalism is denoted $f(t)$. For each s, the level of technology that would have obtained at time t if society had undergone a communist revolution at time s, is denoted $f_s(t)$. That is, the function f_s

traces the profile of technical progress in a counterfactual communist economy taking off at s. The point of time at which our description of the economy starts may arbitrarily be set equal to 0. (This time may, for example, be the year 1750.) Hence the entities $f(t)$ and $f_s(t)$ are defined for non-negative values of s and t. Of course, $f_s(t)$ is defined only for $t \geq s$.

I have not defined the concept 'level of technology' precisely. I have in mind something like per capita aggregate production, ignoring the problems of computing such an aggregate. This implies that growth of productivity is measured in absolute terms. Alternatively, the functions f and f_s could measure the logarithm of per capita production; this corresponds to considering relative growth. The discussion below is independent of the exact definition of $f(t)$ and $f_s(t)$, although of course the interpretation of some of the axioms and results may depend on this definition.

I then state the axioms and comment briefly on them.

I. The function f is continuous. In other words, I assume that the productive forces under capitalism develop smoothly; there is never a sudden jump in the level of technology, neither upwards nor downwards.

II (a). For all t and all $\triangle t > 0$, $f(t+\triangle t) > f(t)$

II (b). For all t, t' and $\triangle t$ with $t > t'$ and $\triangle t > 0$,

$f(t+\triangle t) - f(t) > f(t'+\triangle t) - f(t')$

Part (a) is the axiom of continuous technical progress under capitalism, i.e. a postulate that stagnation never occurs. Part (b) is the axiom of increasing rate of technical change. (If f is twice differentiable, $f'(t)$ measures the rate of change in the level of technology at time t, and $f''(t)$ measures the rate of change in the rate of change. In this case, Axiom II is equivalent to $f'(t) > 0$ and $f''(t) > 0$ for all t. For our purpose, there is no reason to assume differentiability.)

III. For each s, f_s is continuous. This corresponds to Axiom I, except that we now consider the development under communism. While Axiom III says that $f_s(t)$, for a fixed s, is continuous in t, we can also formulate the following postulate:

III'. For each t, $f_s(t)$ is a continuous function of s. The meaning is the following: Let t be a point of time in the future, and consider the hypothetical levels of technology under communism, for different times s of the revolution. As s varies, the axiom requires that $f_s(t)$ vary smoothly and not make sudden jumps. Most of the results can be proved without invoking this axiom; whenever it is used, this fact is pointed out.

IV. For all s, $f_s(s) = f(s)$. This is also a kind of continuity assumption. It requires that the communist economy should start at the same level of technology as the capitalist economy it replaces. If technology is seen as technical knowledge, this assumption appears plausible; it is less so if technology is seen as capital goods embodying that knowledge. I shall not here explore the variants of the axiom that would take into account the possibility of an initial setback due to sabotage, destruction of machinery, and interruption of production.

V. For all s, s', t, t' and $\triangle$t with t $\geqslant$ s, t' $\geqslant$ s' and $\triangle$ t $\geqslant$ 0: If $f_s(t) = f_{s'}(t')$, then $f_s(t + \triangle t) = f_{s'}(t' + \triangle t)$.

This is a consistency requirement. It should be irrelevant for the future growth of a communist economy whether its present state has been reached by capitalism or by communism branching off from capitalism some time in the past. In other words, if two hypothetical communist economies are at the same level at t and t' respectively, then they follow identical courses from these times onwards. Objections may be raised to this idea. Would not socialist man, as he emerges from capitalism, show 'hysteresis traces' (Georgescu-Roegen 1971, p. 126) from his past? As readers of the *Critique of the Gotha Program* will know, the question is an extremely important one, but in order not to overburden the model I shall disregard it here.

VI. There exist a time s and a number A such that, for all t $\geqslant$ s, $f_s(t) \leqslant A$. This is the assumption of the initial necessity of capitalism. Axiom II implies that f(t) will ultimately attain any level. If, therefore, there is a time s such that a communist economy taking off at s will forever be bounded from above, then revolution should not occur at s.

VII. There exists a time s such that, for all t > s, $f_s(t) > f(t)$. This is the assumption of the ultimate superiority of communism. If the communist revolution occurs at a sufficiently late date, from that date onwards the level of technology will be consistently higher than it would have been under capitalism. This axiom incorporates the idea that when capitalism has created the conditions that permit universal want satisfaction, then the conditions for spontaneous creative innovation will also be present.

Of these axioms, II follows from the text to be interpreted. Axioms I and III are fairly uncontroversial technical assumptions. Axioms IV and V are simplifying assumptions that permit us to disregard some complexities of real-life revolutions, such as an initial set-back or 'the tradition of all the dead generations [that] weighs like a nightmare on the brains of the living' (Marx 1852, p. 103). The crucial axioms from a substantial point of view are VI and VII. Axiom VII can be weakened in an obvious manner; this is explored in the next section.

It is shown in the Mathematical Note, Section A.1, that the axiom system is consistent.

4. Discussion of the model

The main interest of the model, exegetical precision apart, is to permit a framework for the analysis of the transition from capitalism to communism. I first discuss the normative or welfare-theoretical aspects of this issue: assuming that a revolution will occur if there is a need for it, when *should* it occur? What is the optimal point of transition? I further raise the analytical and strategic question whether it is at all likely that a revolution *will* occur when needed.

There are a number of possible transition points, each corresponding to a particular welfare criterion. The first is

> C_1. The transition should occur at the earliest possible time at which a communist economy branching off could make unlimited technical progress.

To simplify the further discussion, we define a set I by $s \in I$ if and only if for all A there exists a t such that $f_s(t) > A$. That is, I is the set of points in time at which a revolution will lead to unlimited technical progress under communism. Criterion C_1 says that the transition should occur at the earliest possible time in I. The set I has a simple structure: there exists a time S_1 such that no point before S_1 and all points after S_1 belong to I. In fact, we can say something more. The following theorem is proved in the Mathematical Note, Section A.2:

> *Theorem 1.* There exists a time S_1 with the following properties:
> (a) If $s < S_1$, then $s \notin I$ and $f_s(t) \leq f(S_1)$ for all $t \geq s$
> (b) If $s > S_1$, then $s \in I$ and f_s is a strictly increasing function
> (c) If Axiom III' holds, then $S_1 \in I$ and $f_{S_1}(t) = f(S_1)$ for all $t \geq S_1$.

In informal terms, the theorem says that if the revolution occurs before S_1, the communist economy will not make unlimited progress; in particular it will never exceed the technological level reached under capitalism at time S_1. If the revolution occurs after S_1, the communist economy will make uninterrupted and unlimited progress. Under the additional continuity assumption III', we also know the result of a revolution exactly at time S_1. It will lead to a complete and everlasting stagnation; the communist economy will neither grow nor contract.

There is, incidentally, a difficulty in applying C_1. If S_1 does not belong to I (as Theorem 1(c) tells us it will certainly not do if Axiom III' holds), the phrase 'earliest possible time' is not defined. This can be solved by stipulating that the transition point be S_1+d, where d is some arbitrarily chosen small constant. Henceforward we shall ignore this problem. S_1 then is the transition point defined by C_1.

The next criterion to be discussed is

C_2. The transition should occur at the earliest possible time at which the communist economy that branches off will ultimately overtake capitalism, even if initially it lags behind capitalism.

According to the criteria to be discussed below, the transition should not occur until communism can be instantaneously superior, but according to C_2 ultimate superiority is sufficient. According to C_1 there is no need for superiority ever to be achieved; it is sufficient that any level attainable by capitalism can also – even if perhaps much delayed – be attained by communism. The choice between C_1 and C_2 boils down to this: should communism compete with capitalism (C_2) or be content simply to follow in its path, however sluggishly (C_1)? The choice between C_3 (or the closely related C_5) and C_2 will be seen to be this: should the material conditions for communism be developed by capitalism if this is the more rapid way, or should one prefer the slower development whereby communism itself creates the foundations for its own future prosperity?

We then define a set J by the condition $s \in J$ if and only if $f_s(t) > f(t)$ for all $t > s$. The definition is closely related to Axiom VII; the axiom simply says that J is not empty. The following criterion suggests itself:

C_3. The transition shall occur at the earliest possible time in J.

Let S_3 denote the transition time defined by C_3. In formal terms, $S_3 = \inf J$. (See the discussion above of a difficulty in talking about 'the earliest possible time' satisfying a certain condition.) The following theorem is proved in the Mathematical Note, Section A.3:

Theorem 2. Suppose that $s \in J$. For any $s' > s$ and all $t \geq s'$, $f_s(t) > f_{s'}(t)$.

The implication is the following. A communist economy branching off at some point of time in J is not only consistently superior to continued capitalism (as required by the definition of J), but also consistently superior to any communist economy branching off at a later time. In other words, if $s \in J$, there is absolutely no reason, from a welfare-theoretical point of view, to postpone the revolution after s. C_3 is the most conservative criterion we shall consider.

If $s \in J$ and $s' > s$, we might expect $s' \in J$. In the example of Section A.1, this is indeed the case. But it does not follow from the axioms. In fact, the following possibility is consistent with the axioms, as shown by an example in Section A.3:

There exist s and s' with $s \in J$ and $s' > s$ such that, for all $s'' \geq s'$ and all $t > s''$, $f_{s''}(t) < f(t)$.

Although s satisfies criterion C_3, there may exist a time s' later than s such that a communist economy branching off at or after s' will remain forever inferior to capitalism. The revolution should have occurred at s, but if delayed until s' (for whatever reason), it is better to call it off completely. This possibility is not in the best accordance with the theory I am trying to interpret; see the comments to Axiom VII. It is, of course, possible to introduce extra axioms that will rule it out, but I have not found any natural postulates that serve this purpose. Recall, however, that the possibility just described is not a consequence of the axioms, even though it is consistent with them.

The example shows that it is consistent with the model to assume an upper as well as a lower limit to the time at which a communist revolution can successfully occur. This in itself does not seem unreasonable. One might in fact reason as follows: 'The revolution should occur neither too early nor too late; there is a tide in the affairs of men and you have to make the attempt at exactly the right moment (or within exactly the right interval) if it is to succeed.' Now this may be true, but it does not show that the example considered is consistent with the intuitions behind the model. The term 'successful', when applied to the revolution, is ambiguous. It may mean 'successful in establishing communist relations of production', or 'successful in establishing a technical progress that instantaneously or ultimately overtakes capitalism'. The neither-too-early-nor-too-late argument applies to success in the first sense. But *given* success in the first sense, it does not apply to success in the second sense. There are no substantial reasons why postponing the revolution should make for an irrecoverable loss in welfare, even though Theorem 2 says that some loss may be expected. And so the fact that the axioms allow for an upper limit to the time at which communism can branch off and be superior to capitalism shows that they do not completely capture our intuitions. I return below to the relation between the two meanings of 'success'. I now explore some variants of Axiom VII:

VII'. There exists a time s such that (i) for all $t \geq s$, $f_s(t) \geq f(t)$, and (ii) there exists a $t > s$ such that $f_s(t) > f(t)$.
VII". There exists a time s such that, for all $t \geq s$, $f_s(t) \geq f(t)$.

Corresponding to VII' and VII" we can define sets J' and J", criteria C_4 and C_5, and transition times S_4 and S_5. The formal definitions are not given, since they are obvious in view of the discussion of VII, J, C_3 and S_3. Clearly $J \subseteq J' \subseteq J''$ which implies $S_3 \geq S_4 \geq S_5$. In the Mathematical Note, Section A.4, the following is proved:

Theorem 3. $S_4 = S_5$

The theorem is proved under the assumption that Axiom VII holds, though in fact Axiom VII' will suffice. But if only VII" holds, it is possible that J' is empty while J" is not, in which case Theorem 3 is false. We would then have $f_s(t) = f(t)$ for all s and t with $t \geq s > S_5$; that is, after S_5, a revolution has no effect on the level of technology. This is hardly consistent with the theory that I am interpreting; therefore I rule it out by sticking to Axiom VII.

Theorem 3 shows that VII' and VII" define only one new criterion, which can be thus formulated in words:

C_5. The transition should occur at the earliest possible time at which there is a communist economy branching off that never becomes inferior to capitalism.

Could all the criteria coincide? If the continuity assumption III′ is imposed, they will not. It is proved in the Mathematical Note, Section A.4:

Theorem 4. Suppose that III′ holds. Then $S_1 < S_5$.

The choice between C_3 and C_5 boils down to this: is the technical level of communism an *argument* for bringing about that system, or only a *condition* for doing so? According to C_5, superior technical efficiency is not the reason for bringing about communism, but nevertheless this criterion makes efficiency count for more than C_1. It is more difficult to compare C_2 and C_5. In a sense C_5 insists more on efficiency, since it says that communism must be non-inferior *from its very beginning,* but in another and perhaps more important sense C_2 implies a heavier stress on efficiency, since it says that communism must ultimately become *superior.* C_5 says that inferiority at any time is an argument against communism, C_2 that ultimate superiority is an argument for it. And of course, C_3 implies a heavier stress on efficiency than either C_2 or C_5.

I believe that C_1 gives too little weight to technical efficiency. According to C_1, any delay in raising standards of living would be justified, as long as one could be sure that ultimately they would be raised. At the other end of the spectrum, C_3 certainly gives too much weight to efficiency. I believe that this is also the case for C_5, for one should be prepared to accept at least some time during which communism is inferior to capitalism before overtaking it (and one should be prepared to do this independently of the problems that are assumed out of existence by Axioms IV and V). Still this does not mean that I embrace C_2, for if the roundabout method of overtaking capitalism takes too long a time, it would be unethical to impose this sacrifice on a large number of generations. These remarks point to an obvious weakness in the model: its excessively ordinal character. For an adequate discussion of the problem one would want to know *how much better* one mode of production performs than the other at any given time, or *how long time* it takes for one mode to overtake the other. Nevertheless I believe that some insight

CONTRADICTION BETWEEN FORCES AND RELATIONS OF PRODUCTION 225

into the nature of the problem can be gained from a purely ordinal model, as in other cases.

I now turn to a discussion of the realism of the model. I do *not* discuss the realism of the basic assumption VII, derived from Marx's philosophical theory. It should be said, however, that the obvious failure of the actual communist countries to overtake capitalism, or even to avoid being outpaced, is not relevant here. In the model this can be interpreted simply by saying that the revolution in these countries occurred too early, according to one or more (possibly all) of the above criteria. Nor do I discuss the evidence that would be needed in order to form some notion of the shape of the functions f and f_s. Nor, finally, do I enter upon a further discussion of the simplifying assumptions IV and V. Instead I want to focus upon the *political realism* of the model. Is it at all likely that this 'contradiction between forces and relations of production' could bring about a political revolution?

To see the problem in its proper context, let us observe that the theory of contradiction between capitalist forces and relations of production is but one of several theories of capitalist crises proposed by Marx. There is, above all, the theory of the falling rate of profit; there are proto-Keynesian elements of a theory of over-saving: and vague notions about disproportionality crises, overproduction, and so on. To evaluate these theories, observe that, in addition to empirical validity and logical consistency, a Marxist theory of crises should possess three main features. (i) It should be *dialectical,* in the sense that I have explored elsewhere (Elster 1978a, Ch. 5). (ii) It should be *materialist,* in the sense of invoking irreversible changes. (iii) It should explain how the crises provide a *motivation to revolutionary action.* It is clear that all theories that make the crises depend on technical change satisfy the second condition. A demand crisis of the Keynesian variety is not irreversible; the multiplier works both ways, and government intervention may restore the full-employment equilibrium. (This is not to say that the resolution of the crisis may not have some consequences that are hard to reverse, such as the establishment of a machinery for government intervention.) By contrast the fall in the rate of profit, according to Marx, is irreversible because linked to technical progress that cannot be undone once it has occurred. The fact that the theory of the falling rate of profit is shot through with logical holes (Elster 1978a, pp. 113ff.) does not affect this methodological point. Nor does it detract from the exemplary dialectical structure of the theory. The fall in the rate of profit comes about by the very actions undertaken to counteract it; these actions are individually rational for the entrepreneurs but collectively disastrous. This, incidentally, also is true of the Keynesian theories. Finally, both the theory of the falling rate of profit and the Keynesian theory are linked to action, but only the former to revolutionary action, because of the irreversible nature of the contradiction.

What about the contradiction between productive forces and relations of production, seen in the light of these desiderata? As observed above, the theory certainly is materialist. It is not, perhaps, dialectical in the rigorous sense proposed by Elster (1978 a), but certainly in the looser sense of a system that by its very success undermines its own existence. The great weakness of the theory, however, is that it is very hard to link it to action. The counterfactual nature of the base-line for evaluating capitalist performance makes the theory too abstract to serve as a basis for action. If one could point to a declining performance over time, this might be an incentive to change the system, but we have seen that according to Marx the contrary is true. If one could compare the existing capitalist system in one country with an existing communist system in another, then the superior performance of the latter might motivate a change of regime in the former, but this argument could not hold for the first country to make the transition. Here the problem is similar to the one encountered in the comparative study of industrialization. We may be able to explain the industrialization in Germany, France or Russia by invoking the gap between actual and potential performance (Gerschenkron 1966, p. 8), but this is only because the potential was realized elsewhere, viz. in England. It would be absurd to explain the industrialization of England along these lines.

Let me pursue this suggestion somewhat further. The idea is that the superiority of communism would explain the communist revolution in all countries but the first in which it occurred. Here the explanation must be a different one, which need not concern us here. So the appearance of communism on the world historical scene could be more or less accidental, but its subsequent expansion would be rationally grounded. This, for example, would make the present argument compatible with the theory offered in Cohen (1978). But an obvious presupposition of this is that the revolution should not occur too early in the first country. Even if we dismiss the notion that the revolution might occur in the pioneer country *because* communism would be more efficient, it is essential that it should occur at a moment *when* communism would be more efficient, because otherwise there will be no success to inspire the latecomers. 'But societies are not so rational in building that the dates for proleterian dictatorship arrive exactly at that moment when the economic and cultural conditions are ripe for socialism.' (Trotsky 1977, p. 334) In the light of historical experience we might go even further, and suggest – following Trotsky himself – that communist revolutions invariably occur in industrially backward countries, which not only fail to overtake capitalism, but even are outpaced by it. (True, this initial outpacement might be due to the problems that we have discussed in connection with Axioms IV and V above, but the Soviet experience points to a different lesson.)

The tragic implication of this line of argument is the following. For the communist revolution to be victorious, i.e. to create communist relations of production, it must occur so early in the process of industrialization that it cannot be successful in the sense of creating a rate of technical progress satisfying one of the criteria C_2 through C_5 (and possibly not even C_1). In the development of a society there is a period during which it is politically ripe for communism, and a period during which it is economically ripe, and these two periods do not seem to have any overlapping parts. At least this would seem to be a plausible argument for countries that cannot look to a successful instance to imitate, and if it is valid for these countries there will be no instance to imitate. The argument can be broken only by showing that there are other roads to communism than an abrupt and wholesale change in economic relations, a task that cannot be undertaken here.

I have argued until now on the assumption that comparisons with hypothetically superior systems are not politically very effective. The Norwegian Conservative Party in 1961 understandably had little success when *in opposition* they adopted the slogan that the British Conservatives (and later the Danish Social Democrats) used *in power:* 'Make good times better.' Internal crises or external examples may suffice to topple a régime, but not an abstract possibility of a superior way of doing things. Let us imagine, nevertheless, that such abstractions may have motivating power. There would still remain the following problem. Even if the working class in a given country came to see the inherent superiority of communism as stated in Axiom VII, it would be too much to expect that it should also accept the initial necessity of capitalism as stated in Axiom VI. A natural impatience might very easily bring about a premature transition, violating one or more (possibly all) of the criteria for transition. And even if the workers could be persuaded to support capitalism for the required period of time, the better to destroy it later, it might be questioned whether the capitalist class would be very eager to receive support on these premises. The Menshevik argument of helping capitalism to power has in its favour the economic argument that we have been discussing. It can also be backed by some remarks about 'crude communism' in *The Economic and Philosophic Manuscripts* (Marx 1844, pp. 294ff.). But from the strategic and political point of view it is hopelessly out of touch with reality. The class struggle does not align itself nicely to coincide with the 'historical missions' assigned to the different classes by the productive forces.

But, it will be asked, what if the insight into the ultimate superiority of communism emerges at a time when capitalism is no longer necessary? This no doubt was what Marx had in mind. And he may have been right, but nevertheless mistaken in the implicit assumption that the insight would emerge in a country where capitalism no longer was necessary. As explained by Knei-Paz (1977), it was the achievement of Trotsky to show that the most advanced revolutionary consciousness among the workers goes hand in hand with the backward character of capitalist relations in a country. Latecomers to the process of industrialization have a small proportion of the work force in industry, but what workers there are, are concentrated in a few huge factories that embody the most advanced techniques of mass production. Marx was aware of this complex relation between the centre and the periphery of capitalism (Marx 1845, pp. 74–5, 1850, pp. 134–5), but did not draw the full conclusions from his insight. He did not realize, that is, that for a successful communist revolution the conditions must be united *in one country*. It is not sufficient that one country achieves an economic level that is both sufficient for communism to be viable economically in that country and for the communist revolution to be politically viable in another.

Mathematical Note

BY AANUND HYLLAND

A.1. Consistency of the axioms

We prove that the axiom system is consistent by presenting an example of functions f and f_s that satisfy Axioms I–VII. In the example, Axiom III' also holds. Unfortunately, the construction of the function f_s is a little complicated. We have to consider several cases, depending on the value of s. Comments on the definition are given along the way.

(a) $f(t) = t^2$, for all $t \geq 0$.
(b) Let s satisfy $0 \leq s < 1$ and define

$$g(s) = s + \frac{s}{1-s}$$

For these values of s, f_s shall be a non-increasing function. Until time $t = g(s)$, $f_s(t)$ shall be strictly decreasing in t, at which time it shall reach level 0 and stay there forever after. (Here 0 can, for example, represent the technological level of a non-industrialized pre-capitalist society.) The function g is increasing, $g(0) = 0$, and $g(s)$ tends to infinity as s tends to 1 from below.

For s and t with $0 \leq s < 1$ and $s \leq t \leq g(s)$, we shall find the number s' that satisfies

$$g(s') - s' = g(s) - t$$

Straightforward calculations show that s' is unique and is given by

$$s' = 1 - \frac{1}{1 + g(s) - t}$$

It is easy to see that s' is a decreasing function of t. For all possible values of t, we have $0 \leq s' \leq s$, while $t = s$ gives $s' = s$ and $t = g(s)$ gives $s' = 0$. In order that Axioms IV and V be satisfied, we define

$$f_s(t) = f_{s'}(s') = f(s') = s'^2.$$

For $t > g(s)$, we let

$$f_s(t) = 0.$$

Each function f_s is continuous. For any fixed number $t \geq 1$, if s tends to 1 from below, then s' tends to 1, and so does $f_s(t)$.

(c) $f_1(t) = 1$, for all $t \geq 1$.
(d) Then let s satisfy $1 < s \leq 2$, and define

$$h(s) = s - 1 + \frac{1}{s-1}$$

For the values of s now considered, f_s shall be a strictly increasing function that grows without bounds. But for $s < 2$, the growth of $f_s(t)$ shall lag behind the growth of $f(t)$, at least for values of t immediately above s. In particular, only at time $t = h(s)$ shall $f_s(t)$ reach the value $f(2) = 4$. The function h is decreasing; it satisfies $h(2) = 2$ and $h(s) > 2$ for $1 < s < 2$, and h(s) tends to infinity as s tends to 1 from above.

For s and t with $1 < s \leq 2$ and $s \leq t \leq h(s)$, we shall find the number s' that satisfies

$$h(s') - s' = h(s) - t.$$

This number is unique and is given by

$$s' = 1 + \frac{1}{1 + h(s) - t}$$

Here s' is an increasing function of t, $t = s$ gives $s' = s$, $t = h(s)$ gives $s' = 2$, and $s \leq s' \leq 2$ for all possible values of t. Then we define

$$f_s(t) = f_{s'}(s') = f(s') = s'^2.$$

For $t \geq h(s)$, we define

$$f_s(t) = 4 \cdot (t - h(s) + 1)^2.$$

Strictly speaking, we have given two definitions of $f_s(t)$ for $t = h(s)$, but they both give $f_s(h(s)) = 4$. Now it is easy to see that each function f_s is continuous. If s tends to 1 from above, then s' and $f_s(t)$ tend to 1; this holds for any fixed $t > 1$.

(e) If we set $s = 2$ in the definitions above, we get $h(2) = 2$ and $f_2(t) = (2t - 2)^2$, for all $t \geq 2$.

This gives $f_2(2) = 4 = f(2)$, as required by Axiom IV. Moreover, $f_2(t) > f(t)$ for all $t > 2$.

By Axioms IV and V, f_s is now determined for any $s > 2$. For any such s, there exists a unique number t' such that

$$f_s(s) = f(s) = f_2(t').$$

We find t' explicitly:

$$t' = 1 + \frac{s}{2}$$

For $t \geq s > 2$, Axiom V now gives

$$f_s(t) = f_s(s + (t-s)) = f_2(t' + (t-s)) = f_2(t+1-\frac{s}{2}) = (2t-s)^2.$$

This completes the construction of f and f_s. Applying the comments already made, it is not difficult to show that Axioms I–VII and III' are satisfied.

The different entities defined in the text have these values:

$I = (1, \infty)$
$J = J' = J'' = [2, \infty)$
$S_1 = S_2 = 1$
$S_3 = S_4 = S_5 = 2$

Criteria C_1 and C_2 are not satisfied at time 1, but they hold at any later time. Therefore, these criteria require that the transition occur at time 1 + d for some small, positive number d. On the other hand, criteria C_3–C_5 are satisfied at time 2 (and any later time); hence these criteria can be applied directly.

A.2. Theorem 1

In order to prove Theorem 1, we investigate the properties of the set I. First, we prove the following:

(*) If $s > s'$ and $s' \in I$, then $s \in I$.

Let s and s' satisfy the premise of (*). By Axioms II (a) and IV, $f_{s'}(s') < f_s(s)$. We can choose $A = f_s(s)$ in the definition of I and conclude that there exists a t such that $f_{s'}(t) > f_s(s)$. Since $f_{s'}$ is a continuous function (Axiom III) there exists a t' such that $f_{s'}(t') = f_s(s)$. Continuity also implies that $f_{s'}$ is bounded on $[s', t']$; let A_o be an upper bound for the function on this interval. Then let any number A be given. By the definition of I, there exists a t" such that $f_{s'}(t") > \max(A_o, A)$. The construction of A_o gives $t" > t'$. Since $f_{s'}(t') = f_s(s)$, Axiom V implies $f_{s'}(t") = f_s(s + t" - t')$, which gives $f_s(s + t" - t') > A$. The number A was arbitrary; therefore, $s \in I$, and (*) has been proved.

Axiom VI immediately implies that there exists a number s with $s \notin I$. By (*), we cannot have $s' \in I$ for any $s' < s$, hence s is a lower bound for I. Axiom II can be used to prove that f increases without bounds. To be precise, the following can be proved: For any A, there exists a t such that, for all $t' \geq t$, $f(t') > A$. This statement implies that any s satisfying Axiom VII must belong to I. Hence I is a non-empty set which has a lower bound. Therefore, I has a unique infimum or greatest lower bound. Let S_1 be the infimum of I.

We then prove Theorem 1(b). Let $s > S_1$ be given. There must exist an s' satisfying $s' \in I$ and $s' < s$; otherwise, s would be a lower bound for I, contradicting the definition of S_1. Then (*) implies $s \in I$. Suppose that f_s is not a strictly increasing function. That is, there exist t and t' such that $s \leq t < t'$ and $f_s(t) \geq f_s(t')$. We consider two cases: (i) For all $t'' \geq t'$, $f_s(t'') < f_s(t)$. The continuous function f_s has a maximum on the compact set $[s, t']$. Let A be greater than this maximum. Then the value of f_s is less than A everywhere, contradicting the fact that $s \in I$. Hence this case is impossible. (ii) There exists a $t'' \geq t'$ with $f_s(t'') \geq f_s(t)$. Since $f_s(t') \leq f_s(t)$ and f_s is continuous, there must exist a t''' with $t' \leq t''' \leq t''$ and $f_s(t''') = f_s(t)$. We can apply Axiom V with $s = s'$ to conclude that $f_s(t''' + \triangle t) = f_s(t + \triangle t)$ for all $\triangle t \geq 0$. If t_1 and t_2 satisfy $t_1 \geq t$ and $t_2 - t_1 = t''' - t$, we can set $\triangle t = t_1 - t$ and conclude that $f_s(t_1) = f_s(t_2)$. That is, after time t, f_s is a periodical function with period $t''' - t$. Therefore, after time t it can take on no other values than those obtained on the interval $[t, t''']$. But the continuous function f_s has a maximum on the compact interval $[s, t''']$, which must then be a global maximum for the function. As above, the existence of such a maximum contradicts $s \in I$. The assumption that f_s is not strictly increasing has led to a contradiction. The proof of Theorem 1(b) is complete.

Turning to Theorem 1(a), we let $s < S_1$ be given. The definition of S_1 immediately gives $s \notin I$. Suppose that $f_s(t) > f(S_1)$ for some $t \geq s$. We have already observed that f increases without bound. Since f is also continuous, there exists an s' with $f(s') = f_s(t)$. Axiom II (a) gives $s' > S_1$, and Axiom IV gives $f_{s'}(s') = f_s(t)$. By Theorem 1(b), which was proved above, $f_{s'}$ is a strictly increasing and unbounded function. Axiom V implies that the same must hold for f_s from time t on. Hence $s \in I$, contradicting an earlier statement and completing the proof of Theorem 1(a).

In order to prove Theorem 1(c), we assume that Axiom III' holds. For $t = S_1$, Axiom IV gives $f_{S_1}(t) = f(S_1)$. Then choose any $t > S_1$, and consider $f_s(t)$ for $s \leq t$. For $S_1 < s \leq t$, Theorem 1(b), Axiom IV, and Axiom II(a) give $f_s(t) \geq f_s(s) = f(s) > f(S_1)$. Since $f_s(t)$ is continuous in s, this implies $f_{S_1}(t) \geq f(S_1)$. For $s < S_1$, Theorem 1(a) gives $f_s(t) \leq f(S_1)$, and the continuity assumption implies $f_{S_1}(t) = f(S_1)$. By choosing $A > f(S_1)$ in the definition of I, we get $S_1 \notin I$. The proof is complete.

A.3. The set J

In this section, we first prove Theorem 2. Then we give an example in which there exist two points of time s and s' such that s < s', the criterion C_3 is satisfied at s, but none of the criteria $C_2 - C_5$ is satisfied at or after s'.

When discussing the set I in Section A.2, we observed that any s satisfying Axiom VII belongs to I; that is, $J \subseteq I$. Let s, s' and t satisfy s < s' ⩽ t and s∈J. Then s∈I, and the proof of Theorem 1(b) shows that f_s is a strictly increasing function. Axiom IV, Axiom II (a), and the definition of J imply that $f_s(s) = f(s) < f(s') < f_s(s')$. Using the continuity of f_s, we see that there exists a t' with s < t' < s' and $f_s(t') = f(s')$. Then Axioms IV and V give $f_s(t') = f_{s'}(s')$ and $f_s(t' + t-s') = f_{s'}(t)$. Since f_s is increasing and t' + t−s' < t, this gives $f_s(t) > f_{s'}(t)$, which is the conclusion of Theorem 2.

Then we want to construct an example in which the following holds:

There exists s and s' with s∈J and s' > s such that, for all s" ⩾ s' and all t > s", $f_{s''}(t) < f(t)$.

We shall base the construction on the example given in Section A.1. For s ⩽ 4, we define $f_s(t)$ in the same way as in that section. We recall the definition that applies when 2 ⩽ s ⩽ 4:

$$f_s(t) = (2t-s)^2.$$

For s ⩾ 4, we define

$$f_s(t) = (2t - 3 - \frac{4}{s})^2.$$

This is a continuous function in t. It is also continuous in s for all s ⩾ 4. (We have two definitions of $f_4(t)$, but they are easily seen to coincide.) Without much difficulty, we can prove that Axiom V holds. Axiom IV requires the following values of f:

$$f(t) = t^2 \text{ for } 0 \leqslant t \leqslant 4,$$

$$f(t) = (2t-3-\frac{4}{t})^2 \text{ for } t > 4.$$

The function f is continuous for all $t \geq 0$ and differentiable except at $t = 4$. The derivative is increasing, and makes an upwards jump at $t = 4$. This is sufficient to conclude that Axioms I and II hold. Axioms III–VI and III' are implied by comments made above and in Section A.1. Finally, $f_2(t) > f(t)$ for all $t > 2$, establishing Axiom VII.

The various entities defined in the main text have the following values:

$I = (1, \infty)$
$J = J' = J'' = [2,3]$
$S_1 = 1$
$S_2 = \dfrac{3}{2}$
$S_3 = S_4 = S_5 = 2$

Now choose $s = 2$ and $s' = 4$. The definitions immediately give $f_{s''}(t) < f(t)$ for all s'' and t with $s' \leq s'' < t$. The criteria $C_2 - C_5$ are not satisfied at or after time 4. Hence the example satisfies the condition that motivated its construction.

A.4. The relation between the different criteria

Theorems 3 and 4 contain statements about the relation between the different criteria. They are proved in this section.

By definition, $S_4 = \inf J'$ and $S_5 = \inf J''$. Since $J' \subseteq J''$, this gives $S_4 \geq S_5$. If Theorem 3 is false, therefore, we must have $S_4 > S_5$. Then there must exist an s satisfying $s \in J''$ and $s < S_4$. This s belongs to J'' and not to J', which implies $f_s(t) = f(t)$ for all $t \geq s$. Choose any $s' \geq s$. By Axiom IV and the statement just made, we get $f_s(s') = f_{s'}(s')$. Then Axiom V implies $f_s(t) = f_{s'}(t)$ for all $t \geq s'$. The previous observation then gives $f_{s'}(t) = f(t)$ for all $t \geq s'$, which means that s' does not belong to J'. Since s' was an arbitrary number with $s' \geq s$ and since $s < S_4 = \inf J'$, this implies that J' is empty. This, in turn, contradicts Axiom VII (or Axiom VII') and completes the proof of Theorem 3.

To prove Theorem 4, we assume that Axiom III' holds. Choose a number t such that $t > S_1$ and $t > S_5$. For any s with $s \in J''$ and $s \leq t$, we have $f_s(t) \geq f(t)$. By the definition of S_5, there are values of s that belong to J'' and lie arbitrary close to and above S_5. Since $f_s(t) \geq f(t)$ holds for these s, the continuity assumption III' gives $f_{S_5}(t) \geq f(t)$. The choice of t and

Axiom II(a) then imply $f_{S_5}(t) > f(S_1)$. If $S_5 < S_1$, this contradicts Theorem 1(a); if $S_1 = S_5$, it contradicts Theorem 1(c). The only remaining possibility is the conclusion of Theorem 4.

Notes

NOTES TO THE INTRODUCTION TO PART I

1. It is generally accepted that in science there is no 'theory-neutral' observation language. When engaged in deriving observational consequences from a theory to be tested, one always has to take for granted the validity of other theories that enter into the construction of the observation language. The apparently simple notion of a temperature reading embodies a vast amount of theoretical assumptions. But this, in general, is a global difficulty, not a local one, in the sense that rarely does a theory have to be invoked in the construction of the data used for testing that very same theory. We try to construct experiments and observations so that we do not have to presuppose the validity of the theory in order to test it. In some of the social sciences, however, this may not be possible. In linguistics, for instance, the data are our intuitions about grammaticality, which may themselves change as a result of theory changes that make us discover contradictions and connections among the data of which we were formerly unaware, forcing a revision of intuition. There is a distinct possibility that two linguists, working on the same initial set of 'raw intuitions', might go in different theoretical directions and end up with different 'refined intuitions', which correspond perfectly to their theories.
2. The argument for this view is briefly set out in Elster (1978a), p.63, note 8.
3. Elster (1978a, Ch.5) attempts to spell out the structure of social contradictions. In the language of Ch.3 below, they include counterfinality, suboptimality, and games without non-cooperative solutions.
4. Habermas (1978).
5. See the exposition of Peirce's views in Wiener (1949), pp.85ff. For a discussion of the opposite view – that the development of the universe has no absorbing state – in Descartes, see Elster (1975), pp.63–6.
6. Von Wright (1971, Ch.1) has a good account of the Galilean and the Aristotelian traditions in the study of human behaviour. For the underlying physical theories, see Dijksterhuis (1961).
7. Feynman (1965, vol. I, § 26.3) compares the behaviour of light passing from one medium to another to that of a person on land who wants to help a girl drowning in the sea nearby. If the line between him and the girl is not perpendicular to the shoreline, he should not run and then swim in a straight line toward the girl, but rather – if he wants to get to her as soon as possible – follow a broken line which corresponds to that chosen by the light.
8. Samuelson (1971).
9. See Guéroult (1967) and Elster (1975, Ch.5) for some aspects of teleological physics in the seventeenth century.
10. The outstanding cases of such 'as-if' maximization in science are, perhaps, the following. (a) The variational statements of physical theories. These have the feature

that although every process can be described as if something is being maximized, there is no unique objective function that is being maximized in all physical processes. (b) Functional adaptation in biology. Here there is one single maximand in all cases, viz. reproductive capacity. The fact, however, that this is constrained to local maxima only (Ch.2) shows that there is no reason to explain it in terms of maximizing intentions, which are not so constrained (Ch.3). (c) Utility maximization in economics. This is merely shorthand for choice according to consistent, complete and continuous preferences (Elster 1983, Ch.1.2). The artificial or notional character of utility maximization can be seen in many ways. First, there are some rational preferences that cannot be represented by a utility function; secondly even when preferences can be so represented, the representation is not unique; and thirdly the idea of deliberate maximization of utility is somewhat absurd, since utility, like pleasure or happiness, has the property of being *essentially a by-product* (Elster 1983, Ch. 2). It would be possible and interesting to rewrite the history of science in terms of the anthropomorphic notion of maximization, and to view scientific progress as the growing understanding that the idea is purely one of convenience in most cases.
11. Contrary to what is suggested by van Parijs (1981, § 14).
12. See Williams (1966) for a vigorous criticism of this fallacy.
13. Williams (1966, p.209) denounces the idea that 'when one has demonstrated that a certain biological process produces a benefit, one has demonstrated *the* function or at least *a* function of the process'. His examples mainly involve benefits on the group level. And it is not so easy to construct instances of the fallacy at the level of the individual organism. One might consider the case of pleiotropy, and assume that one gene has two phenotypic expressions, both of which are beneficial, but one more so than the other. It is then tempting to say that the least beneficial expression does not explain why the gene is there, since it would be there even in an environment in which that consequence of the gene was somewhat harmful. But it is probably more plausible to refer to this as an instance of functional over-determination, analogous to the well-known case of causal over-determination. See Oster and Wilson (1978, p.314) for comments.
14. Cf. also p. 68 below.
15. Davidson (1980), Chs. 1, 11, 13.
16. Davidson (1980), pp.253-4.
17. This analogy has a family resemblance to one used by Davidson (1980), pp.249-50.
18. Weintraub (1979) has a good survey of this issue.
19. Suppes (1970), p.91.
20. Hicks (1979, Ch.7) has a good discussion of this point.

NOTES TO CHAPTER 1

1. Beauchamp and Rosenberg (1981), p.279.
2. *Ibid.*, p.93.
3. Hempel (1965), pp.335ff.
4. Beauchamp and Rosenberg (1981), p.304.
5. Lewis (1973a).
6. I owe this latter way of phasing the problem to G.A. Cohen. Observe that the phenomenon of preemptive causation differs from that of causal overdetermination, in which each of two actually operating causes is sufficient for the effect, as when a person is hit simultaneously in the heart by two bullets.

NOTES TO CHAPTER 1   239

7. Thorpe (1980, p.68) suggests this distinction.
8. The situation can be compared to the cases where a real process in discrete time is mathematically represented in continuous time, which for computational purposes is then represented in a discrete time that bears no significant relation to the original discrete process (Bellman 1961, pp.67–8).
9. Elster (1975, pp.163ff.) quotes and discusses the relevant texts.
10. Suppes (1970), p.31.
11. *Ibid.*, p.32.
12. See Elster (1976) for a more extensive discussion.
13. McFarland (1970), p.473.
14. See, however, note 49 to Ch.2 below for discussion of the idea that a phenomenon might be explained by its tendency or disposition to produce certain consequences. And this might hold even if the consequences are somehow blocked from occurring.
15. Georgescu-Roegen (1971), p.124, quoting David Bohm and de Broglie.
16. This term (introduced by C.H. Waddington) is the dynamic analogue of the better-known notion of homeostasis. Just as the latter can be illustrated by the idea of a ball rolling in a bowl, the former can be understood by imagining a ball rolling in a valley sloping towards the ocean. If the ball is diverted from its course on the floor of the valley and set rolling up the hillside, then sooner or later it will return to the floor and pursue the course it would have taken in the absence of the diversion – unless the shock was so strong as to send it over into the adjoining valley and make it pursue a different course altogether. The last clause makes the point clear that homeorhesis, like homeostasis, is a concept of *local* stability.
17. Rader (1971). Like many first attempts, this one is flawed by excessive generality, which may account for its sterility in begetting further research.
18. The following draws on Elster (1976).
19. See in particular Georgescu-Roegen (1971). He may be right in insisting on the uniquely historical character of the social sciences, but he must be wrong when suggesting that this leads to hysteresis in some real sense. The historical character of the social sciences may mean that historical processes often are *irreversible*, but this feature can at most distinguish them from physical processes, not from biological evolution. Or it may mean that social phenomena often are *non-repeatable*, e.g. because they have a chaotic and open-ended character that is indissociably linked with their 'firstness'. In society the past survives in the present through *memories* of the past, unlike biological evolution, which is unable to learn from past mistakes or to keep alive past knowledge simply because it may prove useful.
20. See Padulo and Arbib (1974) for an extensive discussion.
21. For three very different evaluations, see Harcourt (1973), Blaug (1974) and Bliss (1975). Since I touch upon this controversy again in the Introduction to Part II and in Ch.7 below, I should briefly state my own (not very competent) evaluation of the issue. First, it appears to me that the controversy in one sense is over: the neoclassical economists lost. Secondly, however, no one yet knows how important this problem in pure theory will turn out to be for applications. And thirdly, there may be, as argued by Blaug (1974, p.82), more that unites Paul Samuelson and Joan Robinson than separates them. In particular, compared to the disequilibrium models of technical change discussed in Ch.5 and Ch.6, the neoclassical and the neo-Ricardian theories may be birds of a feather.
22. 'The ideological role of the "value of capital" is that of breaking the direct actual link between the *time pattern* of labour and the *time pattern* of output in which any technology can be resolved, and establishing instead a relation between *current* output and *current* labour. To this purpose the *current* "value of the capital stock" is needed; a

mythical conceptual reconstruction in which the past and the future of the economy are telescoped into the present.' (Nuti 1970, p.54). This must be wrong. The 'value of the capital stock' can serve the purpose indicated by Nuti; and if chosen to serve this purpose, may be chosen for ideological reasons. But the same purpose can also be served, quite respectably and non-ideologically, by a disaggregated set of current capital goods. And indeed, as argued in the text, this must be a superior representation to the one proposed by Nuti in the opening sentence of the quoted passage.

23. Hicks (1979, Ch.2) rests his whole analysis of causation on the purported counterfactual implication. Cf. in particular the statement on p.14: 'If not-A can be constructed, and with not-A B would not have happened, I shall say that A *passes the test;* granted that A and B exist, A is shown to be a separable cause of B.'
24. Goldman (1972).
25. This is, I believe, the general and as it were pre-Marxist notion of exploitation, as may be confirmed by consulting the OED, Littré and Duden.
26. Roemer (1982).
27. Consider, for instance, the relation between the permanently handicapped and the able-bodied paying them welfare benefits.
28. Stalnaker (1968), Lewis (1973b).
29. Mackie (1962), Beauchamp and Rosenberg (1981), pp.146ff; also Elster (1978a), pp.181ff.
30. This was already recognized by Max Weber (1968), p.276.
31. Elster (1978a), pp.181ff. See also the objections formulated by Barry (1980) and Lukes (1980), with the replies in Elster (1980a, b).
32. Fogel (1964), discussed at greater length in Elster (1978a), pp.204ff.
33. We may note here an interesting feature of Fogel's approach. Since his aim is to prove a thesis, and not to arrive at the best estimate, he can finesse some difficult problems by consistently assuming that the dice are loaded against him, i.e. by playing the Devil's Advocate. If his thesis stands up on the most unfavourable assumptions (so unfavourable, perhaps, that they cannot all be true), then it is certainly also valid on more realistic assumptions, whatever these might turn out to be.
34. Reprinted as Ch.6 in David (1975). The criticism is supported by David's theory of technical change, discussed in Ch.6 below.
35. Elster (1978a, pp.201ff.) has a discussion of this particular problem; see also Kula (1970), p.28.
36. Elster (1980a, b) has a more elaborate argument to this effect.
37. 'There is a magnificence in this generalization which recalls the youth of philosophy. Justice is a perfect cube, said the ancient sage; and rational conduct is a homogeneous function, adds the modern savant.' (Y.C. Edgeworth, quoted after Baumol 1977, p.578, n.11.)
38. Elster (1978a, pp.87ff.) proposes an analysis of the fallacy of composition involved in this approach. See also Elster (1978b, c).
39. Marx (1894), Ch.48; see also the quote from M. Nuti in note 22 above.
40. Elster (1978a, pp.192ff.) discusses the problem of historical alternatives to imperialism as an instance of this difficulty.
41. Russel (1941).
42. This is the illusion that no one would have taken the place of the slaveholders had they not existed. For a brilliant discussion, see Veyne (1976), pp.148ff.
43. For another instance of this problem, cf. the argument in Rawls (1971, pp.177ff and pp.496ff.) to the effect that the least advantaged in a society that makes the worst-off as well off as possible should feel that they are fairly treated. If, in fact, any attempt to improve the situation of the worst-off would make them worse off, the conclusion

appears to follow. But there is a crucial ambiguity here, since the worst-off group under one social arrangement need not be the same as the worst-off under an alternative system, and so a given worst-off group might gain from the change even though the group which then emerges as the worst-off might be worse off than the original worst-off group. This difficulty in Rawls' position was pointed out to me by G.A. Cohen.
44. Here I draw on unpublished work by Trond Petersen.
45. Tocqueville (1969), pp.596–7.
46. *Ibid.*, p.723.
47. *Ibid.*, p.507.
48. *Ibid.*, p.224.
49. *Ibid.*, p.738.
50. Tocqueville (1953), p.343.
51. Here are some examples, with page references to Tocqueville (1969). The process of democratization and the state of democracy induce, respectively, dogmatism and doubt (p.187); hostility and indifference (p.509); loose and strict moral standards (p.599); high and moderate ambitions (p.629); the ruin and the prosperity of industry (p.637).
52. Aron (1967), p.292, Lévi-Strauss (1960), p.94; also Leach (1964), pp.xiff.
53. Hempel (1965), pp.380–1.
54. See Boudon (1977, Ch.IV) for an application of this idea to the choice of career.
55. Castellan (1971), p.51.
56. Hempel (1965), pp.381ff.
57. *Ibid.*, pp.394ff. The example used in the text is taken from Davidson (1980), p.37.
58. Hempel (1965), p.396, commenting on a similar meteorological example.
59. Blau and Duncan (1967), pp.174–5.
60. Simon (1971).

NOTES TO CHAPTER 2

1. The following draws heavily on Elster (1979, Ch.1.2).
2. Frazzetta (1975), p.152.
3. Schumpeter (1934), p.54.
4. Monod (1970).
5. Frazzetta (1975), p.20.
6. Robinson (1956), p.20.
7. Elster (1979, Ch.2) has an extensive discussion of precommitment. See also Thaler and Shefrin (1981).
8. Leigh (1971), Part I; Frazzetta (1975), Ch.5 and *passim*.
9. Cody (1974), Rapoport and Turner (1977), Oster and Wilson (1978), Ch.8.
10. Levins (1968) shows how organisms evolve strategies to cope with seasonal and other regular changes in the environment.
11. For extensions to the case of individuals, see Sen (1967) and Taylor (1976).
12. Williams (1966), pp.212ff.
13. For surveys see Wilson (1975), Ch.4, and Dawkins (1976).
14. Ghiselin (1974) is a frontal attack on this notion.
15. This is a major theme in Dawkins (1976).
16. Elster (1979), Ch.1.5.
17. This, in fact, was the version of the theodicy proposed by Malebranche; see Elster (1975), Ch.5.

18. Surveys are Stark (1962) and Schlanger (1971).
19. One might, perhaps, distinguish between the conservative, the Radical and the Marxist versions of functionalism by the following schematic division. Let us look at the total net product of society as divided into a surplus that goes to the ruling classes and a subsistence component for the rest of society. The conservative functionalism then explains all social arrangements in terms of maximizing the total product (North and Thomas 1973, Posner 1977); Marxist functionalism assumes that the effect is to maximize the first component; and Radical functionalism that the effect is to minimize the second component. To be sure, Radical sociologists (Foucault 1975, Bourdieu 1979) are not mainly concerned with the division of the economic product, but in their work there is much more emphasis on the victims of oppression than on the beneficiaries, and at times one gets the impression that there are in fact no beneficiaries. Since social life is not a zero-sum game, minimizing the welfare of the worst-off may not amount to the same thing as maximizing that of the best-off.
20. Merton (1957), pp.30ff.
21. Cohen (1982) argues, convincingly, that this paradigm is less closely related to evolutionary explanation in biology than I asserted in Elster (1979). He suggests, moreover, that 'Cohen consequence explanations' may be better renderings of the intuitive notion of functional explanation than are 'Elster loops'. But my account is not designed to capture 'intuition' generally, but to make explicit the paradigms used by the foremost defenders and practitioners of functionalist sociology.
22. Of these criteria, (1) through (4) are emphasized by Merton (1957), while criterion (5) is most explicitly stated by Stinchcombe (1968).
23. See Ch.6 for a more detailed analysis of such theories, and for an argument that 'evolutionary models' in economics cannot be slavish imitations of the biological theory of evolution.
24. Hardin (1980).
25. van Parijs (1981), p.101.
16. See Ch.6 for the filter-explanation of technical change found in the work of Eilert Sundt. See also Elster (1979), p.30.
27. Coser (1971), p.60.
28. Marx (1894), pp.600–1.
29. This idea is at the centre of the 'capital logic' school of Marxism, surveyed in Jessop (1977) and in the editorial introduction to Holloway and Picciotta (1978).
30. For a fuller argument, see Elster (1982).
31. This conclusion is drawn in Bowles and Gintis (1977) and Roemer (1979).
32. Simmel (1908), pp.76ff.
33. Filter-explanations, based on the mechanism of deliberate selection among random options, form a subset of the general class of intentional explanations, and one might argue that they have a weaker intentional structure than other members of that class. In particular, intentional selection will not always lead to the global optimum if that option does not come up in the random mechanism, or if it comes up so late in the run that the selector chooses an inferior option that blocks the further development. Yet an intentional selector does have the capacity for waiting or for following indirect strategies, and so is not constrained to local maxima in all cases, contrary to what is suggested by van Parijs (1981), p.56. Thus it could be legitimate to explain phenomena in terms of long-term positive consequences if there is an intentional actor screening the options that are produced by the random mechanism.
34. A classic instance of this argument is Marx's *Eighteenth Brumaire;* a modern version is found in O'Connor (1973), p.70.
35. Thus Leibniz was perfectly entitled, given his theological setting, to explain world

history as an instance of 'reculer pour mieux sauter' (Leibniz 1875–90, vol.III, pp.346, 578; vol.VII, p.568). But without a divine creator and guide, one cannot consistently explain, as did Hegel and Marx, the historical process of alienation through its beneficial consequences for the attainment of a higher unity.
36. In particular, this is how it seems to be understood by Stinchcombe (1974). Cohen (1978, p.258, note 2) thinks differently, but the context is that of a one-shot case (the Hawthorne experiment) in which it is indeed implausible to think that the explanans can succeed the explanandum (Ch.1). Other of Merton's examples, however, concern ongoing activities such as conspicuous consumption, and here it is much more plausible to see his use of the term 'function' as being explanatory.
37. Stinchcombe (1974); see also Stinchcombe (1980).
38. This example, and the sceptical conclusion it suggests, were pointed out to me by Philippe van Parijs.
39. For the corresponding problem about genetic drift and 'non-Darwinian' evolution in biology, see Kimura (1979).
40. For a brilliant study of traditionalism, see Levenson (1968), vol.I, pp.26ff.
41. This example is discussed in Stinchcombe (1968), pp.82–3 and in Stinchcombe (1980).
42. The mechanism could rely on intentional imitation of a constant model, or the superiority of the practice to all nearby alternatives.
43. Leach (1964).
44. Cf. the related argument concerning equilibrium causation in Ch.1 above.
45. Cohen (1978, 1982).
46. Tocqueville (1969), p.605.
47. Veyne (1976), pp.292–3. For a discussion of the occasional functionalism in Veyne's outstanding work, see Elster (1980d).
48. See Cohen (1978), pp.259ff. for a more precise statement.
49. Cohen (1982) argues that feedback loops are not crucial for the causal mechanism by virtue of which a functional explanation, if valid, is valid. Rather, he argues, a consequence law of the form specified above has explanatory power because (or if) A has a propensity or disposition to bring about B. It is, say, a dispositional fact about giraffes at an early stage in their evolution that, were they to develop longer necks, their fitness would be enhanced. To be precise, it is a dispositional feature of the environment of the giraffes that, were giraffes to develop longer necks in that environment, their fitness would increase. Moreover, the environment causes the necks to lengthen by virtue of this dispositional feature. Against this I tend to agree with Ruben (1980) that a constant environment cannot act as a cause, except in the sense of simultaneous causation discussed in Ch.1 above. The issue, however, is in need of further clarification.
50. Festinger (1957, 1964), quoted in Veyne (1976), p.311 and *passim*.
51. In unpublished work on Hempel's philosophy of scientific explanation.
52. Cf. note 11 to the Introduction to Part I.

NOTES TO CHAPTER 3

1. Von Wright (1971) presents a recent statement of the traditional distinction between *Erklären* and *Verstehen,* corresponding roughly to causal and intentional explanation respectively. There are two reasons for not using this terminology. First, it is awkward in that it suggests – contrary to common usage – that we cannot be said to explain something when we understand it, or vice versa. Secondly, and much more importantly, we really need a trichotomy and not a dichotomy. Functional explanation also

enables us to find a *meaning* in the kind of entities it explains, although a less full-blooded meaning than in intentional behaviour. With the dichotomy of understanding vs. explanation, we could say unambiguously that a conscious and intentional act is covered by the first and a convulsion by the second, but we would not know how to classify reflex behaviour such as shutting one's eyes against strong sunlight.
2. For the use of the term 'cause' in this connection, see the comments on the relation between causal and intentional explanation in the Introduction to Part I above.
3. Davidson (1980), Ch.4.
4. Jacob (1970), p.313.
5. Crook (1980), pp.124ff. For a survey of deferred gratification, see Ainslie (1975).
6. Elster (1979, Ch.2) and Ainslie (1980) are surveys of this issue.
7. Elster (1979), pp.13ff. If correct, this observation would imply that intentionality and language are not as inseparable as is argued by philosophers so different as Davidson and Gadamer.
8. Kolm (1980, pp.317–18) sketches a formal model of this idea.
9. Cf. Veyne (1976), p.40 for a similar observation.
10. Elster (1983, Ch.5) has a discussion of this notion.
11. Davidson (1980, Ch.2) and Rorty (1980) are good accounts of this phenomenon.
12. I owe this felicitous phrase to Amos Tversky.
13. Segré (1980), p.171.
14. For a fuller exposition of this criterion (due to Jaakko Hintikka), see Elster (1978a), pp.81ff.
15. Elster (1983, Ch.3) offers a survey of these phenomena.
16. Simon (1954, 1978); see also Ch.6 below.
17. Von Weizsäcker (1965).
18. Heal (1973, Ch.13) has a useful discussion of the existence of optimal plans.
19. Hammond and Mirrlees (1973) introduced this notion.
20. Latsis (1976).
21. The main case in which it is not possible to represent preferences by a real-valued utility function is when they involve a hierarchy of values, as in the so-called 'lexicographic preferences' (Elster 1979, Ch.3.2). Or, to put it the other way around, preferences can be so represented if there are trade-offs between all the objects over which they are defined, so that less of one object may always be compensated for by more of another, the agent remaining at the same utility level.
22. For increasingly complex expositions, see Luce and Raiffa (1957), Harsanyi (1977) and Owen (1968).
23. See notably Roemer (1982) and Howe and Roemer (1981) for normative uses of n-person cooperative game theory.
24. The paradigm example is the proof by Harsanyi that the 'bargaining outcome' to games without a non-cooperative solution also satisfies the optimality criteria laid down by Nash, and is indeed the only outcome that does so. See Luce and Raiffa (1957, Ch.6) for a simple exposition and Harsanyi (1977) for the full account.
25. This is also the view of Popkin (1979), who successfully refutes the so-called 'moral economy' approach to peasant life.
26. Sen (1967). For comparative analyses of the Prisoners' Dilemma, the Assurance Game, and 'Chicken', see Elster (1981b), Kolm (1981) and van der Veen (1981).
27. The phrase is taken from Taylor (1971), the context being an argument designed to show that solidarity and consensus cannot be explained along the lines of methodological individualism. Rephrasing the argument in game-theoretic terms shows that it rather implies the opposite conclusion (Elster 1982).
28. More generally, one may ask about the probability of the solution being realized when

it is not made up of dominant strategies, a case that also covers all games that have solutions in mixed strategies. For the last problem, see Elster (1978a), p.126.
29. Schelling (1978) has constructed such a framework and applied it to a variety of cases.
30. There is also a third possibility, which may be fairly frequently realized: the outcome is Pareto-suboptimal because there are several collective optima, each of which involves specific advantages to some actors, over and above the general advantages accruing to all actors. Any optimum, that is, would be perceived as unjust by some group which is sufficiently powerful to block it. The strategic structure behind this case can be represented by the game often called 'Battle of the Sexes' (Luce and Raiffa 1957, pp.90ff), in which both sexes prefer to be together rather than to be alone, though one prefers that they be together at a restaurant and the other that they be together at a movie. Some implications for politics are spelled out in Schotter (1981), pp.26ff, pp.43ff.
31. Elster (1982) has a fuller exposition.
32. See Rapoport (1966), pp137ff.
33. Schelling (1963) is at the origin of this notion.
34. Pen (1959, pp.91ff) argues against the tendency to believe that the outcome of bargaining is objectively indeterminate. An instance of this tendency is found in Marx (1894), p.364.
35. These remarks follow Barry (1978).
36. Elster (1979, pp.22-3) has further discussion and references.
37. Maynard-Smith (1973). The present exposition follows Dawkins (1976).
38. Yet both Maynard-Smith (1974) and Dawkins (1976) argue that the model can be understood both as stable polymorphism and in terms of mixed strategies.
39. Elster (1979, Ch.3) has a fuller discussion of these anomalies of rational-choice theory.
40. This way of phrasing the paradox of counterfinality brings out very clearly the link to Sartre's earlier work (Sartre 1943), where he argued that men strive to be simultaneously *en-soi* and *pour-soi*, or tend to treat others as if they possess simultaneously both modes of being. 'Trying to relax' involves trying to be both and a consciousness; commanding another person to be grateful is to ask him to assume these contradictory modes (Elster 1983, Ch.2). In social contradictions of the kind discussed in the text, the inconsistent attitudes found in these psychological contradictions are, as it were, distributed over several persons.
41. Sartre (1960).
42. Elster (1978a, Ch.5) has a fuller account of social contradictions and their role in social theory.
43. Elster (1978a, pp.111ff) has a somewhat fuller exposition.
44. For this notion, see March (1978), Elster (1979, Ch.2), Ainslie (1980), Schelling (1980) and Thaler and Shefrin (1981).
45. For some doubts about this, see Frankfurt (1971) and Elster (1979, Ch.2.9).
46. On the problem of deciding to believe, see Williams (1973), Winters (1979) and Elster (1979, Ch.2.3).
47. Elster (1983), Ch.4.
48. I use the term 'drive' to denote a non-conscious striving or tendency, such as the climbing along a pleasure-gradient referred to earlier. In the present case, we may sometimes be unable to decide whether we are dealing with a drive or a meta-preference, i.e. whether the desires are shaped causally or intentionally. In other cases we are hardly in doubt, as when confronted with 'counter-adaptive preferences', exemplified in a person who always, when in Paris, prefers London to Paris and vice versa. It is hard to see someone deliberately shaping himself so as always to prefer forbidden fruit or think the grass greener on the other side of the fence, though we all

know persons who are in the grip of this perverse drive.
49. Tversky and Kahneman (1981).
50. Nisbett and Ross (1980), p.146.

NOTES TO THE INTRODUCTION TO PART II

1. Thus Roemer (1981) employs extensively the framework of general equilibrium theory that has been widely associated with neoclassical economics. What distinguishes Marxist economic theory are the problems raised and the assumptions made, not the techniques employed in deriving answers to the problems from the assumptions. At least this is the state that ought to obtain, and which Roemer's work does much to bring about.
2. Rosenberg (1976a), Part 3, has extensive discussions of this problem.
3. I owe this way of looking at technology to Per Schreiner.

NOTES TO CHAPTER 4

1. In the more abstract framework used by general equilibrium theory, the production possibilities are conceptualized as a *production set*, which may be any set of n-tuples of commodities (including labour) having positive or negative components and satisfying certain postulates, of which the most important is 'one cannot get something for nothing'. For an exposition, see Takayama (1974), pp.45ff. Although this approach is conceptually superior, it does not seem to lend itself to the analysis of technical *change*.
2. The role of joint production in economic theory is twofold. First, there are some processes which in a perfectly straightforward sense have more than one kind of output, as when sheep provide both wool and mutton. Secondly, we may conceive of the use of fixed capital as a kind of joint production, in which old but still usable machines are seen as an output of the production process, along with the main output or outputs. Joint production certainly is relevant for the theory of technical change (Schefold 1980), but space and competence limits the present discussion to the case of a single output.
3. My reference throughout to the static neoclassical theory is Ferguson (1969).
4. Salter (1960), p.30.
5. Salter (1960), pp.33, 40.
6. Hicks (1932), p.125. This passage is used as an epigraph in Binswanger, Ruttan et al. (1978), which is the most sustained recent defence of the Hicksian view.
7. Salter (1960), pp.43-4.
8. Elster (1978a), pp.118ff.
9. Harsanyi (1977), pp.278ff. See also Elster (1979), pp.120-1.
10. Stinchcombe (1968, p.157) has perceptive remarks on the zone of indifference. Cf. also the comments in Ch.3 above on the task of leadership in bringing about the solution to an Assurance Game.
11. Fellner (1961).
12. Kennedy (1964).
13. Ahmad (1966).
14. Ferguson (1969), p.349.

15. This, indeed, is the fundamental objection raised by Nordhaus (1973) to Kennedy's model.
16. Von Weizsäcker (1973), p.109.
17. Finley (1965) and Genovese (1965) present this argument with respect to ancient and American slavery respectively.
18. North and Thomas (1973), p.62 and *passim*.
19. Elster (1975, Ch.3) has a fuller discussion, with particular emphasis on the late seventeenth century. Leibniz in particular is shown to have been intensely preoccupied with these issues, and to have come fairly close to understanding the need for a patent system.
20. Baumol (1965) provides a good analysis of the State in this perspective.
21. In the present work I do not discuss the possible social or economic causes of scientific progress, except for some brief remarks towards the end of Ch.7. Some tentative ideas are offered in Elster (1975), Ch.1.
22. Schmookler (1966) is the best-known modern theory of demand-induced innovation. Rosenberg (1976a) offers good observations on the relation between the market demand for innovation and the scientific supply of inventions.
23. Rosenberg (1976b) has an interesting discussion of this phenomenon, which has recently acquired increasing importance with the development of ever-faster and ever-cheaper generations of computers.
24. Nordhaus (1969). Later developments (e.g. Kamien and Schwartz 1974) also take account of competition between firms for patents.
25. Dasgupta and Stiglitz (1980a), p.276. The authors confirm theoretically the Schumpeterian correlation between research intensity and industrial concentration, but warn against interpreting it causally.
26. Loury (1979).
27. Loury (1979) observes that the model used by Kamien and Schwartz (1976) fails in this respect, by treating rival behaviour as exogenous. One might question, perhaps, that identical behaviour in equilibrium is a constraint on the solution, but at any rate this is what is standardly assumed.
28. Dasgupta and Stiglitz (1980b).
29. For similar observations in the case of biological equilibrium, see Elster (1979), pp.26ff. The economic and biological problems, however, differ in the following way. In the biological case what must be avoided is to assume that the solution will be realized in cases where it does not consist of dominant strategies. In the economic case one must avoid the assumption that some equilibrium will be realized in games without a solution.

NOTES TO CHAPTER 5

1. Examples of this elusiveness are found in several passages quoted here. The summit, perhaps, of ambiguity is the following passage: 'Things economic and social move by their own momentum and the ensuing situations compel individuals and groups to behave in certain ways whatever they may wish to do – not indeed by destroying their freedom of choice but by shaping the choosing mentalities and by narrowing down the list of possibilities from which to choose.' (Schumpeter 1961, pp.139–40). First, if the situations shape the choosing mentalities, what is the point of saying 'whatever they may wish to'? And secondly, how is 'narrowing the list of possibilities' different from 'destroying the freedom of choice'? Answers might perhaps be given to these questions, but none is to be found in Schumpeter.

2. Reported to me by Raymond Aron.
3. Leontief (1968), p.98.
4. For a comparison between Weber and Schumpeter, see Macdonald (1971). For a comparison between Marx and Schumpeter, see Elliott (1980).
5. Tinbergen (1951), pp.59–60.
6. For a discussion of this issue, see Goodwin (1955); see also Smithies (1951).
7. Schumpeter (1939), p.44.
8. Machlup (1951), p.100.
9. Schumpeter (1934), p.64, note 1.
10. *Ibid.*, p.80.
11. *Ibid.*, p.81, note 1.
12. *Ibid.*, p.91.
13. Elvin (1973). His argument is further developed in Tang (1979).
14. Quoted after Tang (1979), p.5.
15. Elvin (1973), p.298.
16. *Ibid.*, pp.314–5.
17. *Ibid.*, p.314.
18. This, for instance, would be the view of North and Thomas (1973).
19. Schumpeter (1934), p.66.
20. Schumpeter (1939), p.223.
21. *Ibid.*, p.284.
22. See Kanbur (1980) for a critical discussion of this view.
23. Schumpeter occasionally uses the term 'induced innovation', but then only of the swarm of secondary innovations that follow in the wake of the primary one. He does mention in passing that 'most innovations are not only labor-saving themselves, but induce adaptations that are also labor-saving' (Schumpeter 1939, p.574), but offers no argument or explanation of this purported fact. As noted in Binswanger, Ruttan et al. (1978, p.118), there does not seem to be 'any literature that discusses whether size and market power could influence the bias (as distinct from the rate) of technical change', which presumably is what a Schumpeterian theory of biased technical change would have to do.
24. Schumpeter (1934), pp.93–4.
25. Schumpeter mentions only that pecuniary success is linked to the second motive, but surely it may also provide a yardstick for the satisfaction of the third.
26. Schumpeter (1951), pp.36, 84, 130, 144–5.
27. *Ibid.*, p.36.
28. Macdonald (1971, p.78) imputes to Marx the view that the entrepreneur is 'motivated by simple greed' of a hedonistic kind. Elliott (1980, p.49) correctly points out that Marx no more than Schumpeter believed that the entrepreneur had purely hedonistic motives.
29. Macdonald (1971), p.79.
30. Schumpeter (1961), p.137.
31. '[My] experience is to the effect that the average businessman always hopes against hope, always thinks that he sees recovery "around the corner", always tries to prepare for it, and that he is forced back each time by hard objective fact which as long as possible he doggedly tries to ignore. The history of the recent world crisis could almost be written in terms of ineffectual attempts to stem the tide, undertaken in a belief, fostered in this case by all the prophets, that business would be "humming" in a few months. This does not mean that businessmen are always optimistic. Far from it.' (Schumpeter 1939, p.141).
32. Schumpeter (1934), p.85. See the comments in Kanbur (1980).

33. Schumpeter (1961), pp.73–4.
34. Hirschman (1967).
35. Nisbett and Ross (1980), p.271.
36. Assuming, however, the falsity of the 'Ortega hypothesis' that second-rate scientists do not contribute anything to progress in science. See the discussion of this hypothesis in Cole and Cole (1973), Ch.8.
37. Goodwin (1955), Frisch (1933).
38. Usher (1951), p.126.
39. Schumpeter (1939), p.73.
40. *Ibid.*, p.273.
41. *Ibid.*, p.768. See also Schumpeter (1961), Ch.10, for a critique of the theory of 'vanishing investment opportunities'.
42. Schumpeter (1939), p.109.
43. *Ibid.*, p.203.
44. *Ibid.*, p.102. See also Hegel (1970), vol.XII, pp.75–6 for a similar distinction between organic and social change.
45. Schumpeter (1939), p.134.
46. *Ibid.*, p.145.
47. *Ibid.*, p.534; see also the comments in Tinbergen (1951).
48. Schumpeter (1939), p.151.
49. *Ibid.*, p.155.
50. Kaldor (1954), esp. pp.68–9.
51. Schumpeter (1939), p.155.
52. Nelson and Winter (1982).
53. The term 'independent' may, however, be inappropriate if, as suggested in Ch.4, market structure and innovation are determined simultaneously.
54. Kamien and Schwartz (1975).
55. Schumpeter (1961), p.83.
56. Leibniz (1875–90), vol.VI, p.237.
57. Schumpeter (1939), p.496 note 1. Later (*ibid.*, p.516) Schumpeter makes a similar point about the incidence of innovation on employment: innovation, being largely labour-saving, tends to reduce employment in the short run and to create it in the long run.
58. Arrow (1971), p.149.
59. Schumpeter (1939), pp.445–6, 775.
60. *Ibid.*, p.419.
61. *Ibid.*, p.496.
62. *Ibid.*, pp. 96, 145.
63. Mason (1951), p.92.
64. Schumpeter (1939), p.109.
65. Schumpeter (1961), p.134.
66. Schumpeter (1939), p.108.
67. Schumpeter (1961), p.242. This view he also imputes to the classical theory of democracy, to which his own is an alternative. He totally neglects, however, that in another classical doctrine democracy was seen as an end in itself, and political discussion as the good life for man.
68. Schumpeter (1961), p.269.
69. *Ibid.*, p.288.
70. *Ibid.*, p.289.
71. *Ibid.*, p.287. Cf. also the distinctly Schumpeterian theory of the 'political business cycle' (Nordhaus 1975).

72. *Ibid.*, p.279.
73. *Ibid.*, p.283.
74. *Ibid.*, p.282.
75. See Ch.2 above, especially note 33.
76. Schumpeter (1961), p.263.
77. Downs (1957) assumes throughout that the self-interested rational preferences of the voters are given independently of – and in fact shape the efforts of – the political parties.
78. Lively (1975), p.38.
79. Schumpeter (1939), p.73.
80. Schumpeter (1961), pp.263, 258.
81. *Ibid.*, pp.259ff.
82. *Ibid.*, p.261, especially note 9.

NOTES TO CHAPTER 6

1. Beck (1980), p.13. He defines tool use as 'the external employment of an unattached environmental object to alter more efficiently the form, position, or condition of another object, another organism, or the user itself when the user holds or carries the tool during or just prior to its use and is responsible for the proper and effective orientation of the tool' (*ibid.*, p.10).
2. *Ibid.*, p.122. Of these functions, the first two are brought together in the 'extractive foraging for embedded foods' which appears to be particularly important from an ecological and evolutionary viewpoint (*ibid.*, pp.190ff).
3. *Ibid.*, p.105.
4. *Ibid.*, p.106.
5. *Ibid.*, p.108.
6. *Ibid.*, p.218.
7. *Ibid.*, pp.243–4.
8. *Ibid.*, p.218.
9. *Ibid.*, pp.203ff.
10. *Ibid.*, p.55.
11. *Ibid.*, p.87.
12. *Ibid.*, p.28.
13. *Ibid.*, p.162.
14. *Ibid.*, pp.18–9.
15. *Ibid.*, p.195.
16. Fagen (1981), p.455.
17. Beck (1980), p.196.
18. Fagen (1981), p.456.
19. *Ibid.*, p.451.
20. *Ibid.*, pp.464–5.
21. See Ch.2 above and van Parijs (1981), pp.79–80.
22. For the distinction between one-step and two-step reductions, see also Elster (1979), Ch.I.
23. Works by Sundt in English are Sundt (1980).
24. Sundt (1862), p.205.
25. *Ibid.*, pp.211–2.
26. I am grateful to Philippe van Parijs for pointing out this ambiguity to me; see also van Parijs (1981), pp.187ff.

27. Kaldor (1954), p.53.
28. Nordhaus and Tobin (1972), p.2, quoted after Nelson and Winter (1975), p.889.
29. Nelson and Winter (1978), p.526.
30. There are strong and explicit references to Schumpeter in Nelson and Winter (1975, 1977a, 1977b, 1978, 1982).
31. Winter (1964, 1975, 1980).
32. Winter (1964), pp.262-4.
33. Winter (1980), p.16. Similar statements are found in Nelson (1980).
34. Winter (1980), p.18.
35. Winter (1971).
36. Nelson and Winter (1975).
37. Nelson, Winter and Schuette (1976).
38. Solow (1957).
39. Nelson, Winter and Schuette (1976), p.110.
40. *Ibid*.
41. In the model, R stands for required dividends. The rationale for the statement in the text is the following: 'The run with the higher value of R will tend to have lower aggregate capital stock, because investment is determined by the profits net of required dividends. Labor demand, and hence the wage rate, will be lower in the high R run. Thus, profitability tests of alternative techniques will favor more labor-intensive techniques in the high R run.' (Nelson, Winter and Schuette 1976, p.101.)
42. *Ibid.*, p.112.
43. Nelson and Winter (1977a, 1978, 1982).
44. Nelson and Winter (1982) embodies this distinction.
45. Strictly speaking, this is a valid measure only if firms are allowed to become extinct, as they are not in the Nelson-Winter models. Nelson and Winter (1978) show how it is possible to convert an actual firm population into one of equal-sized firms with the same 'concentratión index'. Thus, an actual population of 32 firms might have the same degree of concentration as an industry of 14 equal-sized firms.
46. Nelson and Winter (1982), p.126.
47. *Ibid.*, pp.128 ff.
48. *Ibid.*, p.126.
49. Nelson and Winter (1977), pp.88ff. This is but one of the three interrelated puzzles that arise in this model, the other two being that there is a weak connection between productivity performance and research expenditure but a much stronger one between productivity performance and the number of firms (contrary to the Schumpeterian hypothesis, which states that the impact of the number of firms on productivity works through research expenditure), and that both the level of research expense and the return to it are higher in industries of intermediate size ('the internal maximum problem').
50. Nelson and Winter (1977), p.89.
51. Nelson and Winter (1978), p.525.
52. See especially Winter (1964, 1971, 1975). In the last of these articles Winter lists the following problems of natural-selection theories of the firm. (1) *What are the genes?* A theory of natural selection must be supplemented by a theory of genetics that can explain the transmission of behaviour patterns over time. (2) *Actions vs rules of action.* The economic environment does not act upon the behaviour rules, but directly on behaviour. (3) *Entry, exit and dynamics.* Entry and exit conditions can be specified that permit the indefinite persistence of suboptimal behavioural patterns. (4) *Investment policies.* Firms behaving optimally will acquire resources 'with which to expand' (Friedman 1953), but such firms must *actually* expand if the survival of non-optimizers

is to be threatened. (5) *Scale dependence*. The question arises whether an optimizing firm that expands its operations is still an optimizing firm. (6) *Time dependence*. It is difficult for natural selection models to take account of intertemporal optimization, since an evolutionary contest among time-dependent rules is a peculiar thing.
53. Nelson and Winter (1975), p.472.
54. *Ibid.*, p.469.
55. Nelson, Winter and Schuette (1976), p.104, note 18.
56. Nelson and Winter (1977), p.95.
57. David (1975), p.76.
58. Arrow (1962), p.156.
59. Rosenberg (1976a).
60. David (1975), p.64.
61. For typical statements of this kind, see Nelson (1980), p.64, and Rosenberg (1976a), p.64.
62. For a survey of this problem, see Elster (1979), Ch.3.6.
63. Habakkuk (1962).
64. David (1975), pp.81ff.
65. The following quotes from David are all taken from p.76 of his book.

NOTES TO CHAPTER 7

1. Marx (1867), p.326. Marx cites Franklin, and the phrase should not immediately be taken to represent his own view.
2. Marx (1867), p.180.
3. Marx seems to be committed to the propositions (i) that capitalist relations of production are to be explained by the increase in productivity they bring about, and (ii) that they would not have increased productivity unless accompanied by a great deal of suffering. From this it does follow that the suffering can be explained as a causal by-product of an element that is itself teleologically explained. The phenomenon is familiar in biology, where it is referred to as *pleiotropy,* permitting an indirect functional explanation of functionless phenomena. In some contexts, however, one gets the impression that Marx also imputes a direct teleological importance to the sufferings of the proletariat, since without them it would not be motivated to abolish capitalism.
4. Marx (1857), p.240. The phrase is quoted in its context in Appendix 2 below.
5. Marx, like the other classical economists, recognized that production in agriculture took place with variable coefficients of production, permitting smooth substitution.
6. This understanding of Marx is found in Maarek (1975), pp.42-3; Samuelson (1957), although with some misgivings; Morishima (1973), p.12, although it is not clear whether this is intended as exegesis; Brody (1974), p.18, with a rare defence of the economic plausibility of fixed coefficients; Blaug (1978), p.293. Hjalmarsson (1975) argues that Marx held coefficients to be fixed *ex post,* but variable *ex ante,* as in the 'putty-clay' model of Johansen (1959). In my view the texts cited by Hjalmarsson are compatible with this interpretation, but do not require it.
7. The following is taken from Ferguson (1969), Ch.2. He does not, however, consider the idea of imposing condition (2) without condition (3). In fact, given condition (2), condition (3) appears to be quite implausible: if the coefficients depend on the level of output, one would expect them to change in relative as well as absolute terms.
8. Marx (1884), p.27.

9. Marx (1867), p.346.
10. *Ibid.*, p.380.
11. Marx (1894), p.145.
12. *Ibid.*, p.79.
13. *Ibid.*, p.82.
14. Marx (1867), pp.393–4. Here he also quotes Ricardo to the effect that 'Machinery.... can frequently not be employed until labour (he means wages) rises'.
15. *Ibid.*, p.631.
16. *Ibid.*, p.393.
17. Elster (1978b). I am indebted to Robert van der Veen for pointing out to me the implications of the texts referred to in notes 14–16 above.
18. It is often recognized in passing that production permits several kinds of substitution. For instance, around each point on the unit isoquant one may draw a local isoquant of greater curvature, the unit isoquant being the outer envelope of all these local isoquants. We may think of the local isoquants as embodying *ex post* substitution possibilities, and the full unit isoquant as embodying the larger scope for variation that is possible at the drawing board stage. Yet it is not clear whether this representation corresponds to the distinction between intra-technique and inter-technique choice, and in any case it does not have a central place in the literature.
19. For other interpretations, see Elster (1978b).
20. The point, simply stated, is that given the technological possibilities and the wage bundle, we may derive prices and the rate of profit directly, without using labour values as an intermediate step. See again Elster (1978b).
21. Roemer (1981), p.203; also Roemer (1982).
22. See, for instance, Shaikh (1978) and the vigourous rebuttals by Steedman (1980) and van Parijs (1980).
23. Roemer (1981), Ch.3 has a full discussion.
24. Rosenberg (1976a, Ch.7) has a useful survey of Marx's *obiter dicta* on science in its relation to economic growth. It emerges clearly that Marx offers no specific arguments for the view that science is endogenously shaped by economic factors, although he is committed to this view, since he believed that science is an increasingly important part of the productive forces and that the task of the relations of production is to develop the productive forces.
25. Weber (1920), p.203.
26. Marx (1867), p.393.
27. Roemer (1981), Ch.4.
28. Cf. the title of Landes (1969).
29. Elster (1982) argues for a game-theoretical approach to the labour-capital relationship.
30. Marx (1847), p.207. Marx was so struck by the self-acting mule that at times he used it as a symbol for capital itself 'das Kapital als selfactor' (Marx 1861–3, pp.1605, 1620).
31. Marx (1867), pp.435–6. Lazonick (1979) shows that the issue was in fact more complicated.
32. Braverman (1974).
33. Bhaduri (1973).
34. Marglin (1976), p.22.
35. Andrew Ure, quoted in Marglin, (1976), p.30.
36. Marglin, (1976), pp.21–2.
37. Roemer (1981), p.48. This is a fundamental Marxian theorem because it shows that the existence of positive profits presupposes exploitation.
38. Roemer (1981) p.58.

39. *Ibid.*, p.58.
40. *Ibid.*, p.54. Let me give a crude analogy to drive home my point. Different persons will typically educate their children differently. Typically they will also be different when described by a vector of measurable quantities, such as weight, height, blood pressure etc. Yet there is no reason to believe that their methods for child rearing will vary *continuously* with this vector, although it could well be formally possible to set up a function linking education to the vector. The burden of proof is on Roemer to show that his case is not an instance of the same fallacy.
41. Blaug (1968, p.236) quotes several examples of capital-saving innovations during the Industrial Revolution.
42. Dobb (1940), p.125; Sweezy (1942), p.88.
43. Marx (1867), p.638.
44. Marx (1894, p.212) strongly suggests the argument set out in the following.
45. Roemer (1981), Ch. 4–5.
46. Marx (1857), p. 754.
47. In the discussion of Fig.5 it is made clear how to derive the rate of profit in a correct way.
48. Shaikb (1978), refuted in Roemer (1981), Ch.5.
49. Sweezy (1942), p.88.
50. Rowthorn (1980, Ch.7) has an extensive discussion of this issue.
51. Elster (1978a, p.117) has a numerical example exhibiting this possibility.
52. Blaug (1980a), pp.66ff and Blaug (1980b), pp.41ff.
53. Cohen (1978), Ch.8.
54. Kolakowski (1978) has an extensive survey of these epicyclical varieties of historical materialism.
55. This is the central argument in Cohen (1978).
56. Van Parijs (1981).
57. See Ch.2 above for the concept of consequence explanation.
58. Cohen (1978), pp.272–3.
59. Elster (1981b) sketches some objections, acknowledged by Cohen (1982).
60. Brenner (1976).

NOTES TO APPENDIX 1

This paper stems from work done for the Swedish Energy Commission. I have received advice and criticism from Dag Prawitz, Paul Hofset, Gunnar Hals, Dagfinn Føllesdal, Per Schreiner, Steinar Strøm and Jan Døderlein. Improvements have also resulted from the discussion at a meeting of the Working Group on Rationality at the Maison des Sciences de l'Homme, Paris, 4 and 5 December 1978.

1. For a non-specialist it is impossible to master the technical literature on the energy issue. As will be clear from the references, my knowledge about these matters derives from *Science,* which I have read systematically for the last six years or so. I can only hope that I have not missed too many relevant points or taken my cues from work representing peripheral views.
2. Goodin (1978) makes a distinction between modest uncertainty and profound uncertainty, the former being closely related to what I here call risk and the latter arising because we are uncertain about the set of possible outcomes itself. I agree that

profound uncertainty is a widespread phenomenon; indeed it is so widespread that it cancels out in the comparison between the alternatives. I do not believe, therefore, that Goodin is justified in the great importance he attaches to this phenomenon. At any rate it is inconsistent to invoke, as does Goodin, the Arrow-Hurwicz criterion or other criteria for decision under uncertainty, as these criteria all suppose that the set of outcomes is well-defined so that one can pick *the* worst, *the* best etc.

3. Important works are Luce and Raiffa (1957), Raiffa (1968), Keeney and Raiffa (1976). See also Hooker et al. (eds.) 1978. A survey of the standard theory of decision under certainty is Intriligator (1971).

4. Important works are Janis and Mann (1977), Allison (1971), Steinbruner (1974), Cyert and March (1963). The work of Herbert Simon and, following him, James March and Sidney Winter, does not fit easily into the normative-empirical distinction; see Elster (1979), Ch.2.4. and 3.5.

5. If the set of alternatives is 'too large', there may not be a best alternative. It may also be the case that the opportunity set contains increasingly good alternatives converging towards a limit that is not itself a member of the set. For a simple example (Heal 1973, Ch. 13), consider an economy trying to find the rate of saving that maximizes consumption over infinite time. For any rate of saving below 100% it is always possible to find another that gives more consumption, but saving 100% obviously gives zero consumption.

6. Arrow and Hurwicz (1972); see also Luce and Raiffa (1957), Ch.13. In the general case one can only affirm that a decision-maker facing uncertainty can rationally only consider best-consequences and worst-consequences, but there are many ways of combining these extremes into a decision rule. One may look at one end only; or at the other end only; or have one lexicographically prior to the other; or consider a weighted combination.

7. Statement (i) reflects the fact that social actors are often motivated by the rewards that accrue to others, be it through envy or altruism. Statement (ii) reflects a general condition of interdependence that characterizes all social behaviour, and statement (iii) the more particular aspect emphasized by game theory.

8. Of course one might consider strategic games between terrorists and guards, but the energy problem has no special features that make these games different from other military games. Also I refer in the text to the Prisoner's Dilemma, which crops up in this problem as in most others. Perhaps one could use game theory to exploit the problem of *optimal secrecy* of protective measures: too much secrecy reduces the deterrent effect, whereas too little secrecy reduces efficiency.

9. One point should be made. The use of the maximin-criterion (act as if the worst will happen) proposed here is *not* related to the use of the maximin-criterion as a solution concept for zero-sum games. Use of the criterion in such games depends on being certain that one's opponent will also use it, which is the exact opposite of decision-making under uncertainty. The latter is a 'game against nature', not against a strategic other.

10. For readers acquainted with the formal literature on collective choice, this exposition of the basic concepts may seem unfamiliar. Let me briefly indicate, therefore, the links between conditions (i)-(v) in the text and the usual conditions of the impossibility theorems:

	PROCESS	RESULT
FORMAL CONDITIONS OF RATIONALITY	Unrestricted domain Pareto-optimality Irrelevance of independent alternatives	Complete ranking Transitive ranking
SUBSTANTIAL CONDITIONS	*Democracy:* Non-dictatorship Liberalism Non-manipulability Ordinalism (one person, one vote)	*Justice:* Utilitarianism Egalitarianism Rawls etc.

11. Sen (1967) proposes this as a solution to the 'liberal dilemma': even if people have 'nosy' preferences regarding other people's behaviour in private, they could decide not to let these preferences count in the social choice.
12. By permitting ordinal-interpersonal comparisons of welfare one gets the Rawlsian theory; by permitting cardinal-interpersonal comparisons one gets the utilitarian theory. See d'Aspremont and Gevers (1977).
13. In particular I believe that this issue provides a more interesting example of the liberal dilemma (Sen, 1970, Ch.6) than the rather abstract or strained examples often used to illustrate this dilemma. Imagine two communities A and B and three alternatives: to locate a nuclear power plant in A, to locate it in B, and to abstain from building the plant. Naming these alternatives a, b and c respectively, we may imagine the following preference constellation: A prefers b to a (it fears the emission of radioactivity and the possibility of accidents) and a to c (it needs the energy). B prefers c to b (a principled opposition to the building of nuclear plants) and b to a (it needs the employment that will be generated by the plant). If we replace the liberal principle by a 'populist principle', saying that each community must have a veto right over the location of the plant in that community, we at once get Sen's paradox. b is socially preferred to a (both A and B have this preference), a is socially preferred to c (because of A's veto), and c is socially preferred to b (because of B's veto).
14. Thus Harvey Brooks, in his 'Comments' in Stanford Research Institute (1976) writes that 'The largest social cost of nuclear power may be associated with the political reactions to an accident or sabotage incident...Because of the greater sensitivity of the public to catastrophic accidents in comparison with statistical deaths, one should compute the possible costs of a shutdown or partial shutdown of all existing nuclear reactors for a long period following an accident to one reactor'. This argument can be interpreted in two directions, depending on whether it is attributed to the public in general or to the politicians. For a rational public opinion it would be rather strange to include its own future *and unwarranted* reactions among the costs of nuclear power, even if such ahead-looking protection against oneself is not unknown from other domains (Elster 1979, Ch.2). Politicians could, of course, rationally include this element in their deliberations, but then they might have to go against the *ex ante* preferences of the electorate, which is not a very democratic procedure, whether it is dictated by paternalism or by politicians' self-interest. A similar problem ('political synergism' in Brooke's term) crops up in the public's assessment of the radiation

danger after an accident. If, say, the number of deaths from cancer rise from 5% to 6% after an accident, then – depending upon the general psychological climate – either *everyone* dying from cancer could impute their illness to the accident, or *no one* might do so.
15. Hohenemser et al. (1977) point to this problem and raise the question (which they answer in the negative) whether future deaths should be discounted to a present magnitude. If this is not done, then the number of delayed deaths from a single reactor accident might approach infinity.
16. Lehrer (1978) proposes a method for aggregating opinions in such cases. Here each expert is asked not only to evaluate the theories, but also to evaluate the other evaluators, and the overall evaluation can then be determined by a suitable iteration procedure. It remains to be shown whether the scientist behaves in a way such as to make this a useful tool for the present purpose; whether, for example, he is compelled to hold that someone whom he evaluates highly as an evaluator of theories should also be evaluated highly as an evaluator of evaluators.
17. The standard argument (note 6) of using a combination of maximin and maximax criteria would seem to apply, but what does it mean in the present context? Should we, for any given consequence give it the largest of the probabilities that are attached to it by any of the competing theories? Or the smallest? Or a weighted combination? And what if the sum of the probabilities thus defined exceeds 100%?
18. A very good coverage is given by Diamond and Rotschild (eds.) (1978). In spite of the title, this is a volume on risk, not on uncertainty as the term is used here. This fact expresses the now dominant trend in many circles, that Bayesian decision theory and risk analysis can be universally applied, and that there is no such thing as decision under uncertainty. It is only fair to point out, therefore, that the present paper represents an approach that will seem antiquated to many – perhaps most – economists and philosophers. Psychologists, on the other hand, may be more receptive to my argument.

References

Ahmed, A. (1966) 'On the theory of induced innovation', *Economic Journal* 76, 344–57.
Ainslie, G. (1975) 'Specious reward', *Psychological Bulletin* 82, 463–96.
Ainslie, G. (1980) 'A behavioural understanding of the defence mechanism' (Mimeographed).
Allison, G. (1971) *The Essence of Decision*, Boston: Little, Brown.
Aron, R. (1967) *Les Etapes de la Pensée Sociologique*, Paris: Gallimard.
Arrow, K. (1962) 'The economic implications of learning by doing', *Review of Economic Studies* 29, 155–73.
Arrow, K. (1963) *Social Choice and Individual Values*, 2nd ed., New York: Wiley.
Arrow, K. (1971) *Essays in the Theory of Risk-Making*, Amsterdam: North-Holland.
Arrow, K. and Hurwicz, L. (1972) 'An optimality criterion for decision-making under uncertainty' in C.F. Carter and J.L. Ford (eds.) *Uncertainty and Expectation in Economics*, Clifton, N.J.: Kelley.
d'Aspremont, C. and Gevers, L. (1977) 'Equity and the informational basis of collective choice', *Review of Economic Studies* 44, 199–209.
Barry, B. (1978) 'Comment' in S. Benn et al. *Political Participation*, Canberra: Australian National University Press.
Barry, B. (1980) 'Superfox' (review of Elster 1978a). *Political Studies* 28, 136–43.
Baumol, W. (1965) *Welfare Economics and the Theory of the State*, 2nd ed., London: Bell.
Baumol, W. (1977) *Economic Theory and Operations Analysis*, 4th ed., Englewood Cliffs, N.J.: Prentice Hall.
Beauchamp, T.L. and Rosenberg, A. (1981) *Hume and the Problem of Causation*, Oxford: Oxford University Press.
Beck, B. (1980) *Animal Tool Behavior*, New York: Garland.
Bellman, R. (1961) *Adaptive Control Processes*, Princeton: Princeton University Press.
Bhaduri, A. (1973) 'A study in agricultural backwardness under semi-feudalism', *Economic Journal* 83, 120–37.
Binswanger, H.P., Ruttan, V.W. et al. (1978) *Induced Innovation*, Baltimore: Johns Hopkins University Press.
Blau, P. and Duncan, O.D. (1967) *The American Occupational Structure*, New York: Wiley.
Blaug, M. (1968) 'Technical change and Marxian economics' in D. Horowitz (ed.), *Marx and Modern Economics*, pp. 227–43. London: MacGibbon and Kee.
Blaug, M. (1974) *The Cambridge Revolution: Success or Failure?*, London: Institute of Economic Affairs.
Blaug, M. (1978) *Economic Theory in Retrospect*, 3d ed., Cambridge: Cambridge University Press.
Blaug, M. (1980a) *The Methodology of Economics*, Cambridge: Cambridge University Press.

Blaug, M. (1980b) *A Methodological Appraisal of Marxian Economics*, Amsterdam: North-Holland.
Bliss, C.J. (1975) *Capital Theory and the Distribution of Income*. Amsterdam: North-Holland.
Boudon, R. (1973) *Mathematical Structures of Social Mobility*, Amsterdam: Elsevier.
Boudon, R. (1974) *Education, Opportunity and Social Inequality*, New York: Wiley.
Boudon, R. (1977) *Effets Pervers et Ordre Social*. Paris: Presses Universitaires de France.
Bourdieu, P. (1979) *La Distinction:* Paris: Editions de Minuit.
Bowles, S. and Gintis, H. (1977) 'The Marxian theory of value and heterogeneous labour', *Cambridge Journal of Economics* 1.
Braverman, H. (1974) *Labour and Monopoly Capital*, New York: Monthly Review Press.
Brenner (1976) 'Agrarian class structure and economic development in pre-industrial Europe', *Past and Present* 70, 30–74.
Brody, A. (1974) *Prices. Proportions and Planning*, Amsterdam: Elsevier.
Castellan, G. (1971) *Physical Chemistry*, 2nd ed., Reading, Mass.: Addison-Wesley.
Cody, M.L. (1974) 'Optimization in ecology', *Science* 183, 1156–64.
Cohen, G.A. (1978) *Karl Marx's Theory of History: A Defence*, Oxford: Oxford University Press.
Cohen, G.A. (1982) 'Functional explanation, consequence explanation, and Marxism', *Inquiry* 25, 27–56.
Cole J. and Cole, S. (1973) *Social Stratification in Science*, Chicago: University of Chicago Press.
Coser, L. (1971) 'Social conflict and the theory of social change' in C. G. Smith (ed.), *Conflict Resolution: Contributions of the Behavioral Sciences*, Notre Dame, Ind.: University of Notre Dame Press.
Crook, J.H. (1980) *The Evolution of Human Consciousness*. Oxford: Oxford University Press.
Cyert, R.M. and March, J. (1963) *A Behavioral Theory of the Firm*, Englewood Cliffs, N.J.: Prentice Hall.
Dasgupta, P. and Stiglitz, J. (1980a) 'Industrial structure and the nature of innovative activity', *Economic Journal* 90, 266–93.
Dasgupta, P. and Stiglitz, J. (1980b) 'Uncertainty, industrial structure, and the speed of R & D', *Bell Journal of Economics* 11, 1–28.
David, P. (1975) *Technical Choice, Innovation and Economic Growth*, Cambridge: Cambridge University Press.
Davidson, D. (1980) *Essays on Actions and Events*, Oxford: Oxford University Press.
Dawkins, R. (1976) *The Selfish Gene*. New York: Oxford University Press.
Demand and Conservation Panel of the Committee on Nuclear and Alternative Energy Systems (1978) 'US energy demand: some low energy futures', *Science* 200, 142–52.
Diamond, P. and Rotschild, M. (eds.) (1978) *Uncertainty in Economics*, New York: Academic Press.
Dijksterhuis, E.J. (1961) *The Mechanization of the World Picture*, Oxford: Oxford University Press.
Dobb, M. (1940) *Political Economy and Capitalism*, 2nd ed., London: Routledge and Kegan Paul.
Downs, A. (1957) *An Economic Theory of Democracy*, New York: Harper.
Elliott, J. (1980) 'Marx and Schumpeter on capitalism's creative destruction', *Quarterly Journal of Economics* 95, 45–68.
Elster, J. (1975) *Leibniz et la Formation de l'Esprit Capitaliste*, Paris: Aubier-Montaigne.
Elster, J. (1976), 'A note on hysteresis in the social sciences', *Synthese* 33, 371–91.
Elster, J. (1978a), *Logic and Society*, Chichester: Wiley.

Elster, J. (1978b) 'The labor theory of value', *Marxist Perspectives* 1 (3), 70–101.
Elster, J. (1978c) 'Exploring exploitation', *Journal of Peace Research* 15, 3–17.
Elster, J. (1979) *Ulysses and the Sirens*, Cambridge: Cambridge University Press.
Elster, J. (1980a) 'The treatment of counterfactuals: reply to Barry', *Political Studies* 28, 143–7.
Elster, J. (1980b) 'Reply to Comments', *Inquiry* 23, 213–32.
Elster, J. (1980c) 'Cohen on Marx's theory of history', *Political Studies* 28, 121–8.
Elster, J. (1980d) 'Un historien devant l'irrationel: lecture de Paul Veyne', *Social Science Information* 19, 773–804.
Elster, J. (1981a) 'Un marxisme anglais', *Annales: Economies, Sociétés, Civilisations* 36, 745–57.
Elster, J. (1981b) 'Introduction' to Kolm (1981) and van der Veen (1981), *Social Science Information* 20, 287–92.
Elster, J. (1982) 'Marxism, functionalism and game theory', *Theory and Society* 11, 453–582.
Elster, J. (1983) *Sour Grapes*, forthcoming from Cambridge University Press.
Elvin, M. (1973) *Patterns of the Chinese Past*, Stanford: Stanford University Press.
Fagen, R. (1981) *Animal Play Behavior*, Oxford: Oxford University Press.
Feiveson, H., von Hippel, F. and Williams, R. (1979) 'Fission power: an evolutionary strategy', *Science* 293, 330–7.
Feller, W. (1968) *An Introduction to Probability Theory and its Applications*, vol. I, 3rd ed., New York: Wiley.
Fellner, W. (1961) 'Two propositions in the theory of induced innovations', *Economic Journal* 71, 305–8.
Ferguson, F.E. (1969) *The Neoclassical Theory of Production and Distribution*, Cambridge: Cambridge University Press.
Festinger, L. (1957) *A Theory of Cognitive Dissonance*, Stanford: Stanford University Press.
Festinger, L. (1964) *Conflict, Decision and Dissonance*, Stanford: Stanford University Press.
Feynman, R. (1965) *The Feynman Lectures on Physics*, Reading, Mass.: Addison-Wesley.
Finley, M. I. (1965) 'Technical innovation and economic progress in the ancient world', *Economic History Review*, 2nd Series. 18, 29–45.
Føllesdal, D. (1979) 'Some ethical aspects of recombinant DNA-research', *Social Science Information* 18, 401–20.
Fogel, R. (1964) *Railroads and American Economic Growth*, Baltimore: Johns Hopkins Press.
Forrester, J. (1971) *Urban Dynamics*, Cambridge, Mass.: M.I.T. Press.
Foucault, M. (1975) *Surveiller et Punir*, Paris: Gallimard.
Frankfurt, H. (1971) 'Freedom of will and the concept of a person', *Journal of Philosophy* 68, 5–20.
Frazzetta, T.H. (1975) *Complex Adaptations in Evolving Populations*, Sunderland, Mass.: Sinauer.
Friedman, M. (1953) 'The methodology of positive economics' in his *Essays in Positive Economics*, Chicago: Chicago University Press.
Frisch, R. (1933) 'Propagation problems and impulse problems in dynamic economics' in R.A. Gordon and L.R. Klein (eds.), *Readings in Business Cycles*, London: George Allen and Unwin 1966.
Genovese, E. (1965) *The Political Economy of Slavery*, New York: Pantheon.
Georgescu-Roegen, N. (1971) *The Entropy Law and the Economic Process*, Cambridge, Mass.: Harvard University Press.

Gerschenkron, A. (1966) *Economic Backwardness in Historical Perspective*, Cambridge, Mass.: Harvard University Press.
Ghiselin, M. (1974) *The Economy of Nature and the Evolution of Sex*, Berkeley and Los Angeles: University of California Press.
Goldman, A. (1972) 'Toward a theory of social power', *Philosophical Studies* 23, 221–68.
Goodwin, R. (1955) 'A model of cyclical growth' in R. A. Gordon and L.R. Klein (eds.), *Readings in Business Cycles*, London: Allen and Unwin 1966.
Goodwin, R. (1978) 'Uncertainty as an excuse for cheating our children: the case of nuclear waste', *Policy Sciences* 10, 25–43.
Guéroult, M. (1967) *Leibniz, Dynamique et Métaphysique*, Paris: Aubier-Montaigne.
Gunsteren, H. van (1976) *The Quest for Control*, Chichester: Wiley.
Haavelmo, T. (1970) 'Some observations of welfare and economic growth' in W. A. Eltis, M. Scott and N. Wolfe (eds.), *Induction, Growth and Trade: Essays in Honour of Sir Roy Harrod*, Oxford: Oxford University Press.
Habakkuk, H.J. (1962) *American and British Technology in the Nineteenth Century*, Cambridge: Cambridge University Press.
Habermas, J. (1978) *Knowledge and Human Interests*, London: Heinemann.
Hammond, P. and Mirrlees, J. (1973) 'Agreeable plans' in J. Mirrlees and N.H. Stern (eds.), *Models of Economic Growth*, London: Macmillan.
Harcourt, G.C. (1973) *Some Cambridge Controversies in the Theory of Capital*, Cambridge: Cambridge University Press.
Hardin, R. (1980) 'Rationality, irrationality and functional explanation', *Social Science Information* 19, 775–82.
Harsanyi, J. (1977) *Rational Behavior and Bargaining Equilibrium in Games and Social Situations*, Cambridge: Cambridge University Press.
Heal, G. (1973) *The Theory of Economic Planning*, Amsterdam: North-Holland.
Hegel, G.W.F. (1970) *Werke in 20 Bänden*, Frankfurt a.M.: Suhrkamp.
Hempel, C. (1965) *Aspects of Scientific Explanation*, New York: Free Press.
Henry, C. (1974) 'Investment decisions under uncertainty: the "irreversibility effect" ', *American Economic Review* 64, 1006–12.
Hesse, M. (1974) *The Structure of Scientific Inference*, London: Macmillan.
Hicks, J. (1932) *The Theory of Wages*, London: Macmillan.
Hicks, J. (1979) *Causality in Economics*, Oxford: Blackwell.
Hirsch, F. (1976) *Social Limits to Growth*, Cambridge, Mass.: Harvard University Press.
Hirschman, A. (1967) *Development Projects Observed*, Washington: The Brookings Institution.
Hjalmarsson, L. (1975) 'Marx and putty-clay', Memorandum from the Department of Economics, University of Gothenburg (Sweden).
Hohenemser, C., Kasperson, R, and Kates, R. (1977) 'The distrust of nuclear power', *Science* 196, 25–34.
Holdren, J. (1978) 'Fusion energy in context: its fitness for the long term', *Science* 200, 168–80.
Holloway, J. and Picciotta, S. (eds.) (1978) *State and Capital*, London: Edward Arnold.
Hooker C.A., Leach, J.J. and McClennen, E.F. (eds.) (1978) *Foundations and Applications of Decision Theory*, vol. 1: *Theoretical Foundations*, Dordrecht: Reidel.
Howe, R.E. and Roemer, J.E. (1981) 'Rawlsian justice as the core of a game', *American Economic Review* 71, 880–95.
Intriligator, M. (1971) *Mathematical Optimization and Economic Theory*, Englewood Cliffs, N.J.: Prentice Hall.
Jacob, F. (1970) *La Logique du Vivant*, Paris: Gallimard.
Janis, I. and Mann, L. (1977) *Decision Making*, New York: Free Press.

Jessop, B. (1977) 'Recent theories of the capitalist state', *Cambridge Journal of Economics* 1, 353–74.
Johansen, L. (1959) 'Substitution *versus* fixed production coefficients in the theory of economic growth', *Econometrica* 27, 157–76.
Kaldor, N. (1954) 'The relation of economic growth and cyclical fluctuations', *Economic Journal* 64, 53–71.
Kamien, M. and Schwartz, N. (1974) 'Patent life and R & D rivalry', *American Economic Review* 183–7.
Kamien, M. and Schwartz, N. (1975) 'Market structure and innovation: a survey', *Journal of Economic Literature* 13, 1–37.
Kamien, M. and Schwartz, N. (1976) 'On the degree of rivalry for maximum innovative activity', *Quarterly Journal of Economics* 90, 245–60.
Kanbur, S.M. (1980) 'A note on risk taking, entrepreneurship and Schumpeter', *History of Political Economy* 12, 489–98.
Keeney, R.L. and Raiffa, H. (1976) *Decisions with Multiple Objectives*, New York: Wiley.
Kelly, J. (1978) *Arrow Impossibility Theorems*. New York: Academic Press.
Kennedy, C. (1964) 'Induced bias in innovation and the theory of distribution', *Economic Journal* 74, 541–7.
Knei-Paz, B. (1977) *The Social and Political Thought of Leon Trotsky*, Oxford: Oxford University Press.
Kimura, M. (1979) 'The molecular theory of neutral evolution', *Scientific American* 241 (5).
Kolakowski, L. (1978) *Main Currents of Marxism*, vols. I–III. Oxford: Oxford University Press.
Kolm, S.-C. (1979) 'La philosophie bouddhiste et les "hommes économiques" ', *Social Science Information* 18, 489–588.
Kolm, S.-C. (1980) 'Psychanalyse et théorie des choix', *Social Science Information* 19, 269–340.
Kolm, S.-C. (1981) 'Altruismes et efficacités: le sophisme de Rousseau', *Social Science Information* 20, 293–344.
Kubo, A.S. and Rose, D.J. (1973) 'The disposal of nuclear waste', *Science* 182, 1205–11.
Kula, W. (1970) *Théorie Economique du Système Féodal*, Paris: Flammarion.
Landes, D. (1969) *The Unbound Prometheus*, Cambridge: Cambridge University Press.
Latsis, S. (1976) 'A research programme in economics' in S. Latsis (ed.), *Methods and Appraisal in Economics*, Cambridge: Cambridge University Press.
Laudan, L. (1977) *Progress and its Problems*, Berkeley and Los Angeles: University of California Press.
Lazonick, W. (1979) 'Industrial relations and technical change: the case of the self-acting mule', *Cambridge Journal of Economics* 3, 231–62.
Leach, E. (1964) *Political Systems of Highland Burma*, 2nd ed., London: Bell.
Lehrer, K. (1978) 'Consensus and comparison: a theory of social rationality' in C.A. Hooker, J.J. Leach, and E.F. McClennen, (eds.), *Foundations and Applications of Decision Theory*, vol. I, Dordrecht: Reidel.
Leibniz, G.W. (1875–90) *Die philosophische Schriften*, ed. Gerhardt, 7 vols. Berlin: Weidmannsche Buchhandlung.
Leigh, E.G. (1971) *Adaptation and Diversity*, San Francisco: Freeman and Cooper.
Leontief, W. (1968) 'The significance of Marxian economics for present-day economic theory' in D. Horowitz (ed.), *Marx and Modern Economics*, London: MacGibbon and Kee.
Levenson, J. (1968) *Confucian China and its Modern Fate*, Berkeley and Los Angeles: University of California Press.

Lévi-Strauss, C. (1960) *La Pensée Sauvage*, Paris: Plon.
Levins, R. (1968) *Evolution in Changing Environments*, Princeton: Princeton University Press.
Lewis, D. (1973a) 'Causation' in E. Sosa (ed.), *Causation and Counterfactuals*, Oxford: Oxford University Press.
Lewis, D. (1973b) *Counterfactuals*, Oxford: Blackwell.
Lindbeck, A. (1976) 'Stabilization policy in open economies with endogenous politicians', *American Economic Review: Papers and Proceedings* 66, 1–19.
Lively, J. (1975) *Democracy*, Oxford: Blackwell.
Loury, G. (1979) 'Market structure and innovation', *Quarterly Journal of Economics* 93, 395–410.
Luce, R.D. and Raiffa, H. (1957) *Games and Decisions*, New York: Wiley.
Lukes, S. (1980) 'Elster on counterfactuals', *Inquiry* 23, 145–56.
Maarek, G. (1975) *Introduction au* Capital *de Karl Marx*, Paris: Calmann-Levy.
McBride, J.P. et al. (1978) 'Radiological impact of airborne effluents of coal and nuclear plants', *Science* 202, 1045–50.
Macdonald, R. (1971) 'Schumpeter and Max Weber: central visions and social theories' in O. Kilby (ed.), *Entrepreneurship and Economic Development*, New York: Free Press.
McFarland, D. (1970) 'Intragenerational social mobility as a Markov process', *American Sociological Review* 35, 463–76.
Machlup, F. (1951) 'Schumpeter's economic Methodology' in S.E. Harris (ed.), *Schumpeter, Social Scientist*, Cambridge, Mass.: Harvard University Press.
Mackie, J.L. (1962) 'Counterfactuals and causal laws' in R. Butler (ed.), *Analytical Philosophy*, First Series, Oxford: Blackwell.
March, J. (1978) 'Bounded rationality, ambiguity and the engineering of choice', *Bell Journal of Economics* 9, 587–608.
Marglin, S. (1976) 'What do bosses do?' in A. Gorz (ed.), *The Division of Labour*, London: Longman.
Mark, R.K. and Stuart-Alexander, D.E. (1977) 'Disasters as a necessary part of cost-benefit analysis', *Science* 197, 1160–2.
Marsily, G.de et al. (1977) 'Nuclear waste disposal: can the geologist guarantee isolation?', *Science* 197, 519–27.
Marx, K. (1844) *Economic and Philosophical Manuscripts of 1844*, in Marx and Engels, *Collected Works*, London: Lawrence and Wishart, vol. III.
Marx, K. (1845) *The German Ideology*, in Marx and Engels, *Collected Works*, London: Lawrence and Wishart, vol. VI.
Marx, K. (1847) *The Poverty of Philosophy*, in Marx and Engels *Collected* Works, London: Lawrence and Wishart, vol. VII.
Marx, K. (1850) *The Class Struggles in France*, in Marx and Engels, *Collected Works*, London: Lawrence and Wishart, vol. X.
Marx, K. (1852) *The Eighteenth Brumaire of Louis Napoleon* in Marx and Engels, *Collected Works*, London: Lawrence and Wishart, vol. XI.
Marx, K. (1853) 'Russian policy against Turkey. – Chartism', in Marx and Engels, *Collected Works*, London: Lawrence and Wishart, vol. XII.
Marx, K. (1857) *Grundrisse*, tr. Nicolaus, Harmondsworth: Pelican.
Marx, K. (1859) *Kritik der politischen Ökonomie*, in *Marx-Engels Werke*, Berlin: Dietz, vol. XIII.
Marx, K. (1861–63) *Zur Kritik der politischen Ökonomie*, in *Marx-Engels Gesamtausgabe*, Zweite Abteilung, vol. III, Berlin: Dietz.
Marx, K. (1867) *Capital I*, New York: International Publishers.
Marx, K. (1884) *Capital II*, New York: International Publishers.

Marx, K. (1894) *Capital III*, New York: International Publishers.
Mason, E.S. (1951) 'Schumpeter on monopoly and the large firm' in S. Harris (ed.), *Schumpeter, Social Scientist*, Cambridge, Mass.: Harvard University Press.
Maynard-Smith, J. (1973) 'The logic of animal conflict', *Nature* 246, 15–18.
Maynard-Smith, J. (1974) 'The theory of games and the evolution of animal conflict', *Journal of Theoretical Biology* 47, 209–21.
Merton, R. (1957) *Social Theory and Social Structure*, Glencoe, Ill.: The Free Press.
Monod, J. (1970) *Le Hasard et la Nécessité*, Paris: Seuil.
Morishima, M. (1973) *Marx's Economics*, Cambridge: Cambridge University Press.
Nelson, R. (1980) 'Production sets, technological knowledge and R & D: fragile and overworked constructs for analysis of productivity', *American Economic Review: Papers and Proceedings* 70, 62–7.
Nelson, R. and Winter, S. (1975) 'Factor price changes and factor substitution in an evolutionary model', *Bell Journal of Economics* 6, 466–86.
Nelson, R. and Winter, S. (1977a) 'Dynamic competition and technical progress' in B. Balassa and R. Nelson (eds.), *Economic Progress, Private Values and Public Policy: Essays in Honor of William Fellner*, Amsterdam: North-Holland.
Nelson, R. and Winter, S. (1977b) 'Simulation of Schumpeterian competition', *American Economic Review: Papers and Proceedings* 67, 271–6.
Nelson, R. and Winter, S. (1978) 'Forces generating and limiting competition under Schumpeterian conditions', *Bell Journal of Economics* 9, 524–48.
Nelson, R. and Winter, S. (1982) 'The Schumpeterian trade-off revisited' *American Economic Review* 72, 114–32.
Nelson, R., Winter, S. and Schuette, H. (1976) 'Technical change in an evolutionary model', *Quarterly Journal of Economics* 90, 90–118.
Nisbett, R, and Ross, L. (1980) *Human Inference: Strategies and Shortcomings of Social Judgment*, Englewood Cliffs, N.J.: Prentice Hall.
Nordhaus, W. (1969) *Invention, Growth and Welfare: A Theoretical Treatment of Technological Change*, Cambridge, Mass.: M.I.T. Press.
Nordhaus, W. (1973) 'Some sceptical thoughts on the theory of induced innovations', *Quarterly Journal of Economics* 87, 208–19.
Nordhaus, W. (1975) 'The political business cycle', *Review of Economic Studies* 42, 169–90.
Nordhaus, W. and Tobin, J. (1972) 'Is growth obsolete?' in R. Gordon (ed.), *Economic Research: Retrospect and Prospect, Economic Growth*, New York: National Bureau of Economic Research.
North, D. and Thomas, R. (1973) *The Rise of the Western World*, Cambridge: Cambridge University Press.
Nuti, D.M. (1970) 'Capitalism, socialism and steady growth', *Economic Journal* 80, 32–57.
O'Connor, J. (1973) *The Fiscal Crisis of the State*, New York: St. Martin's Press.
Olson, M. (1965) *The Logic of Collective Action*, Cambridge, Mass.: Harvard University Press.
Oster, G.F. and Wilson, E.O. (1978) *Caste and Ecology in the Social Insects*, Princeton: Princeton University Press.
Owen, G. (1968) *Game Theory*, Philadelphia: Saunders.
Padulo, L. and Arbib, M. (1974) *System Theory*, Philadelphia: Saunders.
Parijs, P. van (1980) 'The falling-rate-of-profit theory of crisis: a rational reconstruction by way of obituary', *Review of Radical Political Economics* 12, 1–16.
Parijs, P. van (1981) *Evolutionary Explanation in the Social Sciences*, Totowa, N.J.: Rowmand and Littlefield.
Parkins, W.E. (1978) 'Engineering limitations of fusion power plants', *Science* 199, 1403–

8.
Pen, J. (1959) *The Wage Rate under Collective Bargaining*, Cambridge, Mass.: Harvard University Press.
Plamenatz, J. (1954) *German Marxism and Russian Communism*, London: Longman.
Popkin, S. (1979) *The Rational Peasant*, Berkeley and Los Angeles: University of California Press.
Popper, K. (1957) *The Poverty of Historicism*, London: Routledge and Kegan Paul.
Posner, R. (1977) *Economic Analysis of the Law*, 2nd ed., Boston: Little, Brown.
Rader, T. (1971) *The Economics of Feudalism*, New York: Gordon and Breach.
Raiffa, H. (1968) *Decision Theory*, Reading, Mass.: Addison-Wesley.
Rapoport, A. (1966) *Two-Person Game Theory*, Ann Arbor: University of Michigan Press.
Rapport, D.J. and Turner, J.E. (1977) 'Economic models in ecology', *Science* 195, 367–73.
Rawls, J. (1971) *A Theory of Justice*, Cambridge, Mass.: Harvard University Press.
Robinson, J. (1956) *The Accumulation of Capital*, London: Macmillan.
Rochlin, G.I. (1977) 'Nuclear waste disposal', *Science* 295, 23–31.
Roemer, J. (1979) 'Divide and conquer: microfoundations of a Marxian theory of wage discrimination', *Bell Journal of Economics* 10, 695–705.
Roemer, J. (1981) *Analytical Foundations of Marxian Economic Theory*, Cambridge: Cambridge University Press.
Roemer, J. (1982) *A General Theory of Exploitation and Class*, Cambridge, Mass.: Harvard University Press.
Rorty, A. (1980) '*Akrasia* and conflict', *Inquiry* 23, 193–212.
Rosenberg, N. (1976a) *Perspectives on Technology*, Cambridge: Cambridge University Press.
Rosenberg, N. (1976b) 'On technological expectations', *Economic Journal* 86, 525–35.
Roumasset, J. (1976) *Rice and Risk*, Amsterdam: North-Holland.
Rowthorn, B. (1980) *Capitalism, Conflict and Inflation*, London: Lawrence and Wishart.
Ruben, D.-H. (1980) Review of Cohen (1978) in *British Journal of Political Science* 11, 227–34.
Russel, R. (1941) 'The effects of slavery upon nonslaveholders in the ante-bellum South', reprinted in H.D. Woodman (ed.), *Slavery and the Southern Economy*, New York: Pantheon 1966.
Salter, W.G. (1960) *Productivity and Technical Change*, Cambridge: Cambridge University Press.
Samuelson, P. (1957) 'Wages and interest: A modern dissection of Marxian economic models', *American Economic Review* 48, 884–910.
Samuelson, P. (1971) 'Maximum principles in analytical economics' in *Les Prix Nobel en 1970*, Stockholm: Norstedt.
Sartre, J.-P. (1943) *L'Etre et le Néant*, Paris: Gallimard.
Sartre, J.-P. (1960) *Critique de la Raison Dialectique*, Paris: Gallimard.
Schefold, B. (1980) 'Fixed capital as a joint product and the analysis of accumulation with different forms of technical progress', in L. Pasinetti (ed.), *Essays on the Theory of Joint Production*, New York: Columbia University Press.
Schelling, T.S. (1963) *The Strategy of Conflict*, Cambridge, Mass.: Harvard University Press.
Schelling, T.S. (1978) *Micromotives and Macrobehavior*, New York: Norton.
Schelling, T.S. (1980) 'The intimate contest for self-command', *The Public Interest* 60, 94–118.
Schlanger, J. (1971) *Les Métaphores de l'Organisme*, Paris: Vrin.
Schmookler, J. (1966) *Inventions and Economic Growth*, Cambridge, Mass.: Harvard

University Press.
Schotter, A. (1981) *The Economy Theory of Social Institutions*, Cambridge: Cambridge University Press.
Schumpeter, J. (1934) *The Theory of Economic Development*, Cambridge, Mass.: Harvard University Press.
Schumpeter, J. (1939) *Business Cycles*, New York: McGraw-Hill.
Schumpeter, J. (1951) *Imperialism and Social Classes*, New York: Kelley.
Schumpeter, J. (1961) *Capitalism, Socialism and Democracy*, London: George Allen and Unwin.
Segré, E. (1980) *From X-Rays to Quarks*, San Francisco: Freeman.
Sen, A.K. (1967) 'Isolation, assurance and the rate of discount', *Quarterly Journal of Economics* 80, 112–24.
Sen, A.K. (1970) *Collective Choice and Social Welfare*, San Francisco: Holden-Day.
Sen, A.K. (1970) 'Liberty, unanimity and rights', *Economica* 43, 217–45.
Shaikh, A. (1978) 'Political economy and capitalism: Notes on Dobb's theory of crisis', *Cambridge Journal of Economics* 2, 233–51.
Siegenthaler, U. and Oeschger, H. (1978) 'Predicting future atmospheric carbon dioxide levels', *Science* 199, 388–95.
Simmel, G. (1908) *Soziologie*, Berlin: Duncker und Humblot.
Simon, H. (1954) 'A behavioral theory of rational choice', *Quarterly Journal of Economics* 69, 99–118.
Simon, H. (1971) 'Spurious correlations: A causal model' in H. Blalock (ed.), *Causal Models in the Social Sciences*, London: Macmillan.
Simon, H. (1978) 'On how to decide what to do', *Bell Journal of Economics* 9, 494–507.
Slovic, P., Fischhoff, B. and Lichtenstein, S. (1977) 'Cognitive processes and social risk-taking' in H. Jungermann and G. de Zeeuw (eds.), *Decision-making and Change in Human Affairs*, Dordrecht: Reidel.
Smithies, A. (1951) 'Schumpeter and Keynes' in S. Harris (ed.), *Schumpeter, Social Scientist*, Cambridge, Mass.: Harvard University Press.
Solow, R. (1957) 'Technical change and the aggregate production function', *Review of Economics and Statistics* 39, 312–20.
Stalnaker, R. (1968) 'A theory of condititionals' in N. Rescher (ed.), *Studies in Logical Theory*, Oxford: Blackwell.
Stanford Research Institute (1976) 'The economic and social costs of coal and nuclear electric generation', Stanford, California.
Stark, W. (1962) *The Fundamental Forms of Social Thought*, London: Routledge and Kegan Paul.
Steedman, I. (1980) 'A note on the "choice of technique" under capitalism', *Cambridge Journal of Economics* 4, 61–64.
Steinbruner, J.D. (1974) *The Cybernetic Theory of Decision*, Princeton: Princeton University Press.
Stinchcombe, A. (1968) *Constructing Social Theories*, New York: Harcourt, Brace and World.
Stinchcombe, A. (1974) 'Merton's theory of social structure' in L. Coser (ed.), *The Idea of Social Structure: Papers in Honor of Robert Merton*, New York: Harcourt, Brace, Jovanovich.
Stinchcombe, A. (1980) 'Is the Prisoner's Dilemma all of sociology?', *Inquiry* 23, 187–92.
Sundt, E. (1862) 'Nordlandsbåden' in *Verker i Utvalg*, vol. VII, Oslo: Gyldendal 1967.
Sundt, E. (1980) *On Marriage in Norway*, Cambridge: Cambridge University Press.
Suppes, P. (1970) *A Probabilistic Theory of Causality*, Amsterdam: North-Holland.
Swedish Energy Commission (1978) *Miljöeffekter och Risker vid Utnyttjande av Energi*,

Stockholm: Liber Förlag.
Sweezy, P. (1942) *The Theory of Capitalist Development*, London: Dennis Dobson.
Takayama, A. (1974) *Mathematical Economics*, Hinsdale, Ill.: Dryden Press.
Tang, A. (1979) 'China's agricultural legacy', *Economic Development and Cultural Change* 28, 1–22.
Taylor, C. (1971) 'Interpretation and the sciences of man', *Review of Metaphysics* 25, 3–51.
Taylor, M. (1976) *Anarchy and Cooperation*, Chichester: Wiley.
Thaler, R.H. and Shefrin, H.M. (1981) 'An economic theory of self-control', *Journal of Political Economy* 89, 392–406.
Thorpe, J. (1980) *Free Will*, London: Routledge and Kegan Paul.
Tinbergen, J. (1951) 'Schumpeter and quantitative research in economics' in S. Harris (ed.), *Schumpeter, Social Scientist*, Cambridge, Mass.: Harvard University Press.
Tocqueville, A. (1953) *L'Ancien Régime et la Révolution*, vol. II, Paris: Gallimard (Edition des Oeuvres Complètes).
Tocqueville, A. (1969) *Democracy in America*, New York: Anchor Books.
Trotsky, L. (1977) *History of the Russian Revolution*, London: Pluto Press.
Tversky, A. and Kahneman, D. (1974) 'Judgment under uncertainty', *Science* 185, 1124–30.
Tversky, A. and Kahneman, D. (1981) 'The framing of decisions and the rationality of choice', *Science* 211, 453–58.
Usher, A.P. (1951) 'Historical implications of the theory of economic development' in *Schumpeter, Social Scientist*, Cambridge, Mass.: Harvard University Press.
Veen, R. van der (1981) 'Meta-rankings and social optimality', *Social Science Information* 20, 345–74.
Veyne, P. (1976) *Le Pain et le Cirque*, Paris: Seuil.
Weber, M. (1920) 'Die protestantische Ethik under der Geist des Kapitalismus' in *Gesammelte Aufsätze zur Religionssoziologie*, vol. I, Tübingen: Mohr.
Weber, M. (1968) 'Objektive Möglichkeit und adäquate Verursachung in der historischen Kausalbetrachtung' in *Gesammelte Aufsätze zur Wissenschaftslehre*, Tübingen: Mohr.
Weintraub, E. (1979) *Microfoundations*. Cambridge: Cambridge University Press.
Weizsäcker, C.C. von (1965) 'Existence of optimal programmes of accumulation for an infinite time horizon', *Review of Economic Studies* 32.
Weizsäcker, C.C. von (1971) 'Notes on endogenous change of taste', *Journal of Economic Theory* 3, 345–72.
Weizsäcker, C.C. von (1973) 'Notes on endogenous growth of productivity' in J.A. Mirrlees and N.N. Stern (eds.), *Models of Economic Growth*, London: Macmillan.
Wiener, P. (1949) *Evolution and the Founders of Pragmatism*, Cambridge, Mass.: Harvard University Press.
Williams, G.C. (1966) *Adaptation and Natural Selection*, Princeton: Princeton University Press.
Williams, B.A.O. (1973) 'Deciding to believe' in *Problems of the Self*, Cambridge: Cambridge University Press.
Wilson, E.O. (1975) *Sociobiology*, Cambridge, Mass.: Harvard University Press.
Winter, S. (1964) 'Economic "natural selection" and the theory of the firm', *Yale Economic Essays* 4, 225–72.
Winter, S. (1971) 'Satisficing, selection and the innovating remnant', *Quarterly Journal of Economics* 85, 237–61.
Winter, S. (1975) 'Optimization and evolution' in R.H. Day and T. Groves (eds.), *Adaptive Economic Models*, New York: Academic Press.
Winter, S. (1980) 'An essay on the theory of production', forthcoming in the proceedings of the Centennial Symposium of the University of Michigan. Department of Economics.

Winters, B. (1979) 'Willing to believe', *Journal of Philosophy* 76, 243–56.
Wright, G.H. von (1971) *Explanation and Understanding,* Ithaca, N.Y.: Cornell University Press.

Index

Ahmed, A. 104
Arkwright, Richard 173
Aron, Raymond 56
Arrow, Kenneth J. 124, 150, 215

Barton, John 164
Beauchamp, Tom L. 25
Beck, Benjamin 131–5
Bhaduri, Amid 172
Blake, William 24, 206
Blaug, Mark 181
Bohr, Niels 72–3
Braverman, Harry 171

Cohen, G. A. 61, 64, 66–4, 181–4, 192, 210–11, 214, 226
Coser, Lewis 59

Darwin, Charles 56, 66–7, 136
Dasgupta, Partha 109
David, Paul 38, 92, 150–7, 192
Davidson, Donald 22–3
Descartes, René 28
Dobb, Maurice 177
Downs, A. 113

Elster, Jon 189, 190, 192, 199, 201, 216, 225, 226
Elvin, Mark 115, 116

Fagen, Robert 131–5
Feiveson, H. 194

Feller, W. 196
Fellner, William 103, 104, 150
Festinger, L. 67
Føllesdal, D. 198
Fogel, Robert 38, 39, 153
Forrester, J. 202
Frisch, Ragnar 120, 147

Georgescu-Roegen, N. 157, 219
Gerschenkron, A. 226
Goldman, Alvin 35, 36
Goodwin, R. M. 120, 147
Gunsteren, H. van 189

Haavelmo, T. 191
Habakkuk, H. J. 154
Habermas, Jürgen 16
Hardin, Russell 58
Harsanyi, John 103
Hegel, G. W. F. 26, 73, 121-2
Hempel, Carl 26, 29, 44, 46, 68
Henry, Claude 205
Hesse, M. 197
Hicks, John 101, 177–80
Hirsch, F. 191
Hirschman, Albert 119
Hohenemser, C. 196
Holdren, J. 188
Hume, David 25, 30

Jacob, François 71
Janis, I. 199

Kahneman, D. 86, 199
Kaldor, N. 123
Kelly, J. 188
Kennedy, Charles 104–5, 150, 155
Keynes, J. M. 111, 113, 225–6
Knei-Paz, B. 228
Kolakowski, L. 211
Kubo, A. S. 203, 206

Lamarck, J. 66
Leach, Edmund 64
Leibniz, G. W. 28, 56, 124
Leontief, Wassily 95, 112
Lewis, David 26
Lindbeck, A. 190
Loury, Glenn 109, 110
Lyttleton, Humphrey 9

McBride, J. P. 193
Machlup, Fritz 113
Malinowski, Bronislaw 57
Malthus, Thomas Robert 178
Mann, L. 199
Marglin, Stephen 172–4, 181
Mark, R. K. 194
Marsily, G. de 203
Marx, Karl 10, 59, 60, 92, 95, 113, 117, 127, 158–84, 209–12, 214–16, 225, 228
Merton, Robert 57, 61
Monod, Jacques 50
Morishima, M. 175

Nelson, Richard 92, 130, 135, 138–50, 156, 192
Newton, Isaac 28

Oeschger, H. 197, 206

Parjis, Philippe van 182
Parkins, W. E. 188
Pascal, Blaise 203

Peirce, C. S. 18
Plamenatz, J. 211
Popper, K. 189

Ricardo, David 178
Robbins, Lionel 112
Rochlin, G. I. 206
Roemer, John 35, 36, 168, 172, 174–6, 178, 216
Rose, D. J. 203, 206
Rosenberg, Alexander 25
Rosenberg, Nathan 152, 192
Roumasset, J. 186

Salter, W. E. G. 100, 101
Sartre, J.-P. 84
Savage, L. J. 86
Schuette, H. 142, 147, 149
Schumpeter, Joseph 10, 50, 91, 92, 112 ff, 138, 139, 146, 170, 215
Siegenthaler, U. 197, 206
Simmel, G. 60
Simon, Herbert A. 48, 67, 92, 139, 146
Skinner, B. F. 58
Slovic, P. 199
Smith, Adam 108
Solow, Robert 143
Stanford Research Institute 194, 199, 202
Stiglitz, Joseph 109
Stinchcombe, Arthur 57, 61, 64, 148
Stuart-Alexander, D. E. 194
Sundt, Eilert 92, 135–8
Swedish Energy Commission 204
Sweezy, Paul 177

Tawney, R. H. 115
Thompson, d'Arcy Wentworth 52
Tinbergen, Jan 122
Tocqueville, Alexis de 42, 43, 66, 124, 149

Trotsky, Leo 227, 228
Tversky, A. 86, 199

U. S. Committee on Nuclear and Alternative Energy Systems. Demand and Conservation Panel 191
Usher, A. P. 120

Veyne, Paul 65–7

Weber, Max 113, 166
Weizäcker, C. C. von 189
Winter, Sidney 92, 130, 135, 138–50, 156, 192

For EU product safety concerns, contact us at Calle de José Abascal, 56–1°, 28003 Madrid, Spain or eugpsr@cambridge.org.

www.ingramcontent.com/pod-product-compliance
Ingram Content Group UK Ltd.
Pitfield, Milton Keynes, MK11 3LW, UK
UKHW010343140625
459647UK00010B/803